Methods, Algorithms and Circuits for Photovoltaic Systems Diagnosis and Control

Methods, Algorithms and Circuits for Photovoltaic Systems Diagnosis and Control

Editors

Giovanni Spagnuolo
Mattia Ricco

MDPI • Basel • Beijing • Wuhan • Barcelona • Belgrade • Manchester • Tokyo • Cluj • Tianjin

Editors
Giovanni Spagnuolo
University of Salerno
Italy

Mattia Ricco
University of Bologna
Italy

Editorial Office
MDPI
St. Alban-Anlage 66
4052 Basel, Switzerland

This is a reprint of articles from the Special Issue published online in the open access journal *Energies* (ISSN 1996-1073) (available at: https://www.mdpi.com/journal/energies/special_issues/Methods_Algorithms_and_Circuits_for_Photovoltaic_Systems_Diagnosis_and_Control).

For citation purposes, cite each article independently as indicated on the article page online and as indicated below:

LastName, A.A.; LastName, B.B.; LastName, C.C. Article Title. *Journal Name* **Year**, *Volume Number*, Page Range.

ISBN 978-3-0365-0540-4 (Hbk)
ISBN 978-3-0365-0541-1 (PDF)

Contents

About the Editors

Giovanni Spagnuolo (Prof.) is Full Professor of Electrical Engineering at the University of Salerno. Since 2016 he is IEEE Fellow. He was in the 2015 Thomson Reuters list of Most Influential Minds. Since January 2011 he has been serving as Editor of the IEEE Journal of Photovoltaics. Since January 2021 he is Associate Editor of the IEEE Open Journal of the Industrial Electronics Society. Since 2017 he is member of the IEEE European Public Policy Initiative Working Group on Energy. He is co-author of five international patents, of two books and of about 80 journal papers, all on renewable energy systems.

Mattia Ricco (Dr.) received the master's degree (cum laude) in electronic engineering from the University of Salerno, Fisciano, Italy, in 2011, and the Ph.D. double degree in electrical and electronic engineering from the University of Cergy-Pontoise, Cergy-Pontoise, France, and in information engineering from the University of Salerno in 2015. From 2015 to 2018, he was a Postdoctoral Research Fellow with the Department of Energy Technology, Aalborg University, Denmark. Since December 2018, he has been a Senior Assistant Professor (Tenure Track) with the Department of Electrical, Electronic, and Information Engineering, University of Bologna, Bologna, Italy. His research interests include photovoltaic systems, identification algorithms for power electronics, field-programmable gate array-based controllers, battery management system, electric vehicle chargers, digital control of modular multilevel converters.

Preface to "Methods, Algorithms and Circuits for Photovoltaic Systems Diagnosis and Control"

This book collects recent and original contributions in the field of diagnosis and control for photovoltaic systems. Novel diagnostic approaches and related power electronics components and control strategies are investigated with the aim to maximize the photovoltaic output power, to increase the whole system efficiency, to detect internal faults and to guarantee service continuity.

Giovanni Spagnuolo, Mattia Ricco

Editors

Article

Automatic Faults Detection of Photovoltaic Farms: solAIr, a Deep Learning-Based System for Thermal Images

Roberto Pierdicca [1,*], Marina Paolanti [2], Andrea Felicetti [2], Fabio Piccinini [1] and Primo Zingaretti [2]

[1] Department of Civil and Building Engineering and Architecture, Università Politecnica delle Marche, 60131 Ancona, Italy; f.piccinini@pm.univpm.it

[2] Department of Information Engineering, Università Politecnica delle Marche, 60131 Ancona, Italy; m.paolanti@univpm.it (M.P.); a.felicetti@pm.univpm.it (A.F.); p.zingaretti@univpm.it (P.Z.)

[*] Correspondence: r.pierdicca@staff.univpm.it

Received: 26 October 2020; Accepted: 5 December 2020; Published: 9 December 2020

Abstract: Renewable energy sources will represent the only alternative to limit fossil fuel usage and pollution. For this reason, photovoltaic (PV) power plants represent one of the main systems adopted to produce clean energy. Monitoring the state of health of a system is fundamental. However, these techniques are time demanding, cause stops to the energy generation, and often require laboratory instrumentation, thus being not cost-effective for frequent inspections. Moreover, PV plants are often located in inaccessible places, making any intervention dangerous. In this paper, we propose solAIr, an artificial intelligence system based on deep learning for anomaly cells detection in photovoltaic images obtained from unmanned aerial vehicles equipped with a thermal infrared sensor. The proposed anomaly cells detection system is based on the mask region-based convolutional neural network (Mask R-CNN) architecture, adopted because it simultaneously performs object detection and instance segmentation, making it useful for the automated inspection task. The proposed system is trained and evaluated on the photovoltaic thermal images dataset, a publicly available dataset collected for this work. Furthermore, the performances of three state-of-art deep neural networks, (DNNs) including UNet, FPNet and LinkNet, are compared and evaluated. Results show the effectiveness and the suitability of the proposed approach in terms of intersection over union (IoU) and the Dice coefficient.

Keywords: unmanned aerial vehicles; photovoltaic cells inspection; deep learning

---

1. Introduction

With the growing demand for a low-consumption economy and thanks to technological advances, photovoltaic (PV) energy generation has become paramount in the production of renewable energy. Renewable energy sources will represent the only alternative to limit fossil fuel usage and pollution. For this reason, PV power plants are one of the main systems adopted to produce clean energy. Huge investments have been allocated by European countries to stimulate the use of so-called clean energy. Indeed, monitoring the state of health of a system is crucial; detecting the degradation of solar panels is the only way to ensure good performance over time. Besides avoiding a waste of energy, the reason for maintaining a correct functional status of a plant is also economic: the degradation of long-term performance and overall reliability of PV plants can drastically reduce expected revenues [1,2].

PV plants are more and more extensive, composed by thousands of modules, potentially affected by the following fault types: optical degradation or faults, electrical mismatches, and non-classified faults [3]. In the last decades, several methods have been developed, spanning electrical diagnostics, statistical inference from monitored control units, shading detection and so on. Commercial monitoring approaches ensure power loss detection in a portion of the PV field, while the accurate localization of faulty modules requires strings' disassembling, visual inspection, and/or electrical characterization. The long-term performance and the overall reliability of the PV modules strictly depend on faults arising during the operational conditions, or that occurred during the transportation and installation [4,5].

An accurate and prompt detection of defects in the PV modules has the task of guaranteeing an adequate duration time and an efficient power generation of the PV modules and therefore a reliable functioning of the PV plants [6]. Operation and Maintenance (O&M) actions are performed to detect faults. O&M techniques are time-demanding, cause stops to the energy generation, and often require laboratory instrumentation, thus being not cost-effective for frequent inspections [7]. Moreover, it should be noted that PV plants are often located in inaccessible places, making any intervention dangerous.

In this regard, a strong contribution was given by the recent diffusion of unmanned aerial vehicles (UAV), equipped with a thermal infrared sensor, making this technique widely accessible and a de-facto standard for PV fields' diagnosis [8]. The inspection of a PV system using a thermal imaging camera allows to identify any malfunctions of the modules as zones with different colors represent different operating temperatures. Infrared thermography (IRT) is very important for the analysis of PV plants since it allows the acquisition of the operating temperature of each module, an important parameter for the performance evaluations. In addition, even with powerful equipment, accelerating the process of detection of these anomalies it is still challenging; in fact, fault detection is actually very time-consuming and error-prone, since it is generally performed with a visual interpretation of the operator. Moreover, the current practice adopted by the majority of PV plant owners is to perform inspections sporadically, with random criteria and without controlling the overall health of the installation. These represent the main motivations behind the proposed approach [9].

Given the above reasons, in this work, solAIr, a fast and accurate anomaly cells detection system, is developed, leveraging recent advances in deep learning. When dealing with the analysis of large image collections, deep learning-based approaches have been demonstrated to be useful compared to the widely used machine learning approaches (e.g., support vector machines, k-nearest neighbor, decision tree, random forest and more) [10]. The use of a deep neural networks (DNNs) allows a complete understanding of the image, guaranteeing greater accuracy and efficiency and discovering multiple levels of data representation. DNNs can extract the characteristics of the image and automatically classify them from a large amount of image data [11,12]. Hence, the proposed anomaly cells detection system is based on the mask region-based CNN (Mask R-CNN) architecture [13]. This work extends a previous one proposed for the classification of anomaly PV images [14]. In the previous work, a classification task was addressed. For each image, the system deduces that at least one anomaly is present in that image. Instead, in this work the detection task is addressed. For each image, the system returns the exact location of the anomalies contained in the same image. The Mask R-CNN approach solves three tasks at the same time: location, segmentation and classification of objects in an image, generating a bounding box, segmentation mask and class. Additionally, the most important aspect is that the R-CNN Mask solves the segmentation task at the instance level, i.e., it generates a result for each object found. SolAIr was trained and evaluated on the photovoltaic thermal images dataset, a public dataset collected for this work. The dataset is an extension of the one published in the previous work [14]. Initially, the dataset included only images of a portion of Tombourke's system. Now, the dataset has been expanded with images of the entire system. In addition, for each image, a mask containing the segmentation of the faulty cells has been added. The thermal dataset is available (http://vrai.dii.univpm.it/content/photovoltaic-thermal-images-dataset) after compiling a request form in which the applicants specify their research purposes. Furthermore, the performances of three state-of-art DNNs, including UNet [15], FPNet [16] and LinkNet [17], are compared and evaluated in this paper.

The main contributions of this paper can be summarized as follows: (i) a system based on deep learning for the anomaly detection and localization of damaged cells in PV thermal images, (ii) a newly annotated dataset that is publicly available for further experiments by the research community and (iii) a comparison of different deep learning methods that can serve as a benchmark for future experiments in the field.

The paper is organized as follows: Section 2 presents an overview of the related works for PV image processing; Section 3 introduces our approach that consists of a UAV-based inspection system (Section 3.1), gives details on the photovoltaic thermal images dataset (Section 3.2) and introduces a DNN-based solution for anomaly cells detection of PV thermal Images (Section 3.3). Section 4 presents the results, and Section 5 discusses the conclusions and future works.

2. Related Works

The latest technological improvements of digital cameras, in combination with affordable costs, made the PV inspection based on optical methods more and more popular. Specifically, electroluminescence (EL) and IRT imaging represent reliable methods for the qualitative characterization of PV modules. In recent years, several companies have developed systems based on EL techniques. Mondragon Assembly developed an EL inspection system equipped with three high-definition cameras, enabling easy identification of different defects, such as micro cracks, dark areas, finger problems, and short-circuits (https://www.mondragon-assembly.com/solar-automation-solutions/solar-manufacturing-equipment/pv-module-testing/el-inspection/). MBJ implemented high-resolution fully automated electroluminescence test systems for integration into production lines of PV panels, cells, modules or strings. Their system uses deep learning methods to ensure reliable automatic error detection (https://www.mbj-solutions.com/en/products/el-inspection-systems). In addition, AEPVI (Aerial PV Inspection) performs PV power plant inspections by using aerial EL testing systems. The evaluation of the images is automated and uses machine learning techniques to categorize module faults (http://www.aepvi.com/). Quantified Energy Labs performs quantitative electroluminescence analysis (QELA) for enabling the use of EL for outdoor applications. On top of QELA algorithms, they develop machine learning and artificial intelligence models to detect and analyze every module in PV plants and identify potential defects that might reduce the performance of the asset (https://qe-labs.com/). However, as stated in [3], EL-based methods present limitations with respect to IRT imaging which, by contrast, appears likely to be more suitable to provide quantitative information. IRT imaging can provide information about the thermal signature and the exact physical location of an occurring fault, indicating the defective cell, group of cells or module (qualitative diagnosis). In turn, such a thermal signature can be used for quantitative diagnosis, by identifying the electrical power output losses of the impacted module, in the form of dissipated heat. Besides this, thermal images can be obtained in a faster way, with cost-effective tools and avoiding the interruption of energy [3].

For all these reasons, a fault detection method based solely on thermal data will be shown in this paper. However, for the sake of completeness, this section provides the readers with the latest achievements in this field.

Rogotis et al. [18] proposed an early defect diagnosis in PV modules exploiting spatio-temporal information derived form thermal images. The approach uses a global thermal image threshold determined combining two threshold techniques. Their approach is efficient and robust to noise and reflections due to the sun or clouds, but it is not able to detect junction boxes when another area of the panel is super-heated.

In [19], the authors propose the use of standard thermal imaging and the Canny edge detection operator to detect PV module failures that cause the hot spot problem. Several field IRT measurements of thermal images were used for the inspection of defective PV modules. Overall, the whole approach was efficient in detecting hot-spot formations diagnosed in particular defective cells in each module that was analyzed. For this method some limitation occurs as an undesirable sensitivity in case of meaningless background objects. Kim et al. [20] also adopted Canny edge operator and image segmentation techniques to process IRT images acquired with a UAV platform. They used an approach to compare the intensity characteristics of the individual polygons in the panel area.

Efficient and improved edge detection techniques are presented in the work of [21], where significant advancements are presented for automated localization of defects.

In [5] the output IRT images derived from an aerial inspection were processed by the method of aero-triangulation that uses photogrammetry and the global positioning system. Even if all occurring failures are correctly detected, this data treatment method is highly time- and resource-consuming. To solve this problem, some optimizations are currently investigated.

An innovative thermal sensor that experimentally localizes heat sources and estimates the peak temperature using machine learning algorithms (ThermoNet) has been introduced by [22]. The combination of the thermal sensor called ThermoMesh and ThermoNet allows the detection of a high-speed high-resolution heat source through the transfer of conductive heat.

In [23], the authors evaluated and implement an automated detection method to inspect a PV plant using a UAV equipped with IRT, whereas in [24] the effectiveness of PV plant detection based on the profiles of temperature was studied. They also used a UAV equipped with an infrared camera that inspects the quality of photovoltaic systems in real operating conditions. The temperature distribution of PV modules allows to detect the defective modules. A useful approach to identify the presence of hot spots in real time was presented in [23]. But this approach was efficient only for the identification of the aforementioned type of defect, and not for other forms of failure.

Algorithms based on artificial neural networks (ANN) have been proposed to detect anomalies in PV modules. In fact, some recent studies have demonstrated that the use of deep learning can improve the defect detection performance in the aerial images of PV modules, thanks to their ability of self-learning, fault tolerance and adaptability [1]. The work of [25] detects three typologies of PV faults (disconnected substrings, hot spots, and disconnected strings) on infrared images acquired by a thermographic camera mounted on a UAV. The images are processed with digital image-processing methods and then are used as samples for training a CNN. They demonstrated that the algorithm was able to detect faults that were not detectable with the image-processing techniques. Telemetry and IRT images were used to detect hot spots in the work of [26]. Their approach is based on a region-based recurrent convolutional neural network, that once trained, is used as a hot spot detector. The work of [27] compared the performance of hotspot detection in the IRT image of PV modules using two approaches. The first is based on the classical technology that uses Hough line transformation and the Canny operator to detect hotspots. The second uses the deep learning model based on Faster-RCNN and transfer learning. With the second they obtained the best results. Close to the approach proposed in this article is the work of Dunderdale et al. [28]. To identify faulty modules, they combined a scale invariant feature transform (SIFT) descriptor with a random forest classifier. Moreover, to evaluate the performance of deep learning models, VGG-16 and MobileNet were implemented. Conversely, our study advances the state of the art, as it performs a segmentation task, with the advantage of identifying the correct location of each fault. Moreover, our approach exploits the thermal raw data. Finally, the results of the tested methods are compared using state-of-the-art metrics. At a glance, the previous solution [14] for the classification of PV damaged images has been improved by applying recent object detection architectures to the casting anomaly cells detection task, namely: Mask R-CNN [13], UNet [15], FPNet [16] and LinkNet [17]. The details of the proposed methods are presented in the following sections.

3. Materials and Methods

The approach presented in [14], i.e., the classification of PV anomaly images, has been extended for the development of the proposed solAIr system. To the best of our knowledge, this is the first available dataset with thermal information specifically annotated for the management of PV plants. Indeed, the available SoA datasets include RBG [29] or electtroluminescence [30] images, but thermal information is neglected. The framework for the anomaly cells detection system, as well as the novel PV thermal image dataset used for evaluation, were comprised of three main components: the UAV-based inspection system, the mask region-based CNN (Mask R-CNN) architecture and the DNN-based solution (see Figure 1). The design of the defect detection system is based on the Mask R-CNN architecture, which was adopted to simultaneously perform object detection and instance segmentation, making it useful for the automated inspection task. Further details on the UAV-based inspection system and the DNN-based solution are given in the following sections with the evaluation metrics adopted for solving this task. Details on the data collection and ground truth labelling are discussed in Section 3.2.

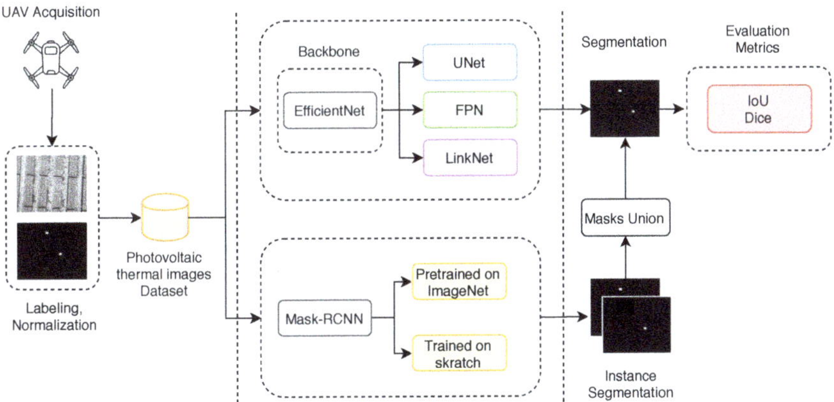

Figure 1. Workflow of the solAIr system—a Unmanned Aerial Veichle (UAV) based intelligent inspection system for monitoring and anomaly cells detection of Photovoltaic Plants (PV). In the first step, a UAV is used to scan the PV system. The acquired frames are annotated and stored in the photovoltaic thermal images dataset. In the next step, the selected neural network (Region Based Convolutional Neural Network -RCNN) is trained on a portion of the dataset. In the last step, the trained models are tested on the remaining portion of the dataset. For the final experimental evaluation, state-of-the-art metrics (like Dice and Intersection over Union (IoU)) are used for the comparison between the segmentation of the networks and the relative ground truth.

3.1. UAV-Based Inspection System

The UAV-based inspection system is based on a Skyrobotic SR-SF6 drone equipped with a radiometric Flir Tau 2 640, a thermal camera with a resolution of 640×512 pixels and a focal length of 13 mm. The detailed UAV specifications and parameters adopted in this work are presented in Table 1. The analysis was carried out with a constant flight altitude of 50 m with respect to the surface of the panels.

The thermalCapture (Tau core) hardware of the thermo-camera can work in different modes; in our case thermal detections are available in two different temperature ranges: "high gain" and "low gain". For "high gain" the range of temperature is between -25 and $+135\,^\circ$C. For "low gain" the range of temperature is between $-40\,^\circ$C and $+550\,^\circ$C, but a lower resolution than the first. All thermo-camera specifications can be found in [14]. Once the raw thermal data are acquired, they can be pre-processed by a thermografic software, in our case ThermoViewer version 3.0.7 (https://thermalcapture.com/thermoviewer/). It is important that the settings in ThermoViewer match those of the Tau Core to provide a valid output of temperature.

Table 1. UAV platform specification and parameters.

Overview	Value
Width (open)	110.0 ± 5 cm
Length (open)	110.0 ± 5 cm
Height	36.0 ± 5 cm
Empty Weight	2250 ± 50 g
Maximum Payload Weight	2000 g
MTOW (Maximum Take Off Weight)	5500 g
Max Endurance	39 min
Storage Temperature Range	-15–70 °C
Operating Temperature Range	-15–50 °C

3.2. Photovoltaic Thermal Images Dataset

In this work, we provide a novel PV thermal image dataset (http://vrai.dii.univpm.it/content/photovoltaic-thermal-images-dataset). For its collection, a thermographic inspection of a ground-based PV system was carried out on a PV plant with a power of approximately 66 MW in Tombourke, South Africa. The thermographic acquisitions were made over seven working days, from 21 to 27 January 2019 with the sky predominantly clear and with maximum irradiation. This situation is optimal to enhance any abnormal behavior of the entire panels or portion thereof.

3.2.1. Dataset Annotation

The images were captured during the inspection of the PV plants. The operator has selected the images with the presence of one or more anomaly cells. Then, the associated binary mask was generated. This mask contains white pixels indicating the anomaly cell. The detection of the anomalous cell is made only through the use of thermal data: the operator immediately identifies where the anomaly is placed because the cell has a temperature value that is totally different from all the surrounding cells. This difference has been evaluated by a software called ThermoViewer (Figure 2).

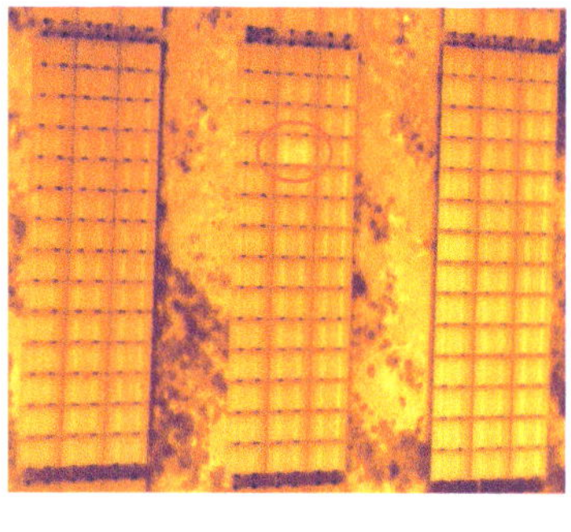

(a)

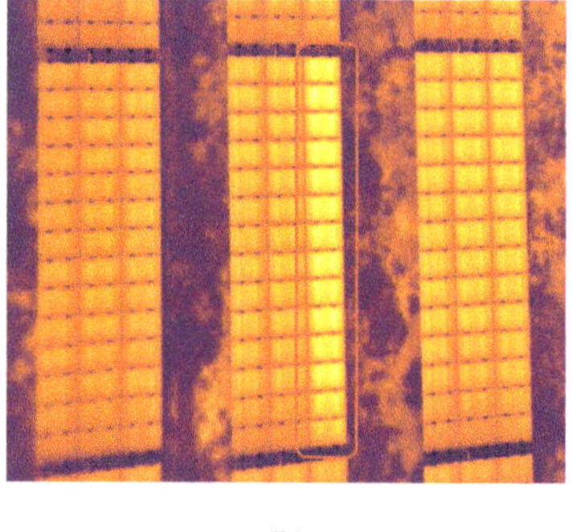

(b)

Figure 2. Examples of raw thermal images showed by ThermoViewer. The anomalous cell is visibly located by the operator, as it has a different temperature range from the surrounding cells. (**a**) A single defected cell; (**b**) a contiguous sequence of faulty cells (string).

The thermal images, obtained with the raw radiometric data, associate a thermal value to each pixel, using the Celsius graduated scale. The images may present one or more anomalies, as depicted in Figure 3, and the operator creates a single mask that segments each anomalous cell. In case of a portion of contiguous anomalous cells, the operator segments the whole portion in a single block. The pre-processing and annotation phase produced a dataset of 1009 thermal images, including each respective mask. The thermal images and the binary masks have the same dimensions of 512×640 pixels.

The input classes were chosen according the following three types of annotation:

- Images with one anomalous cell (Figure 3a,b);
- Images with more than one anomalous cell (Figure 3c,d);
- Images with a contiguous series of anomalous cells (Figure 3e,f).

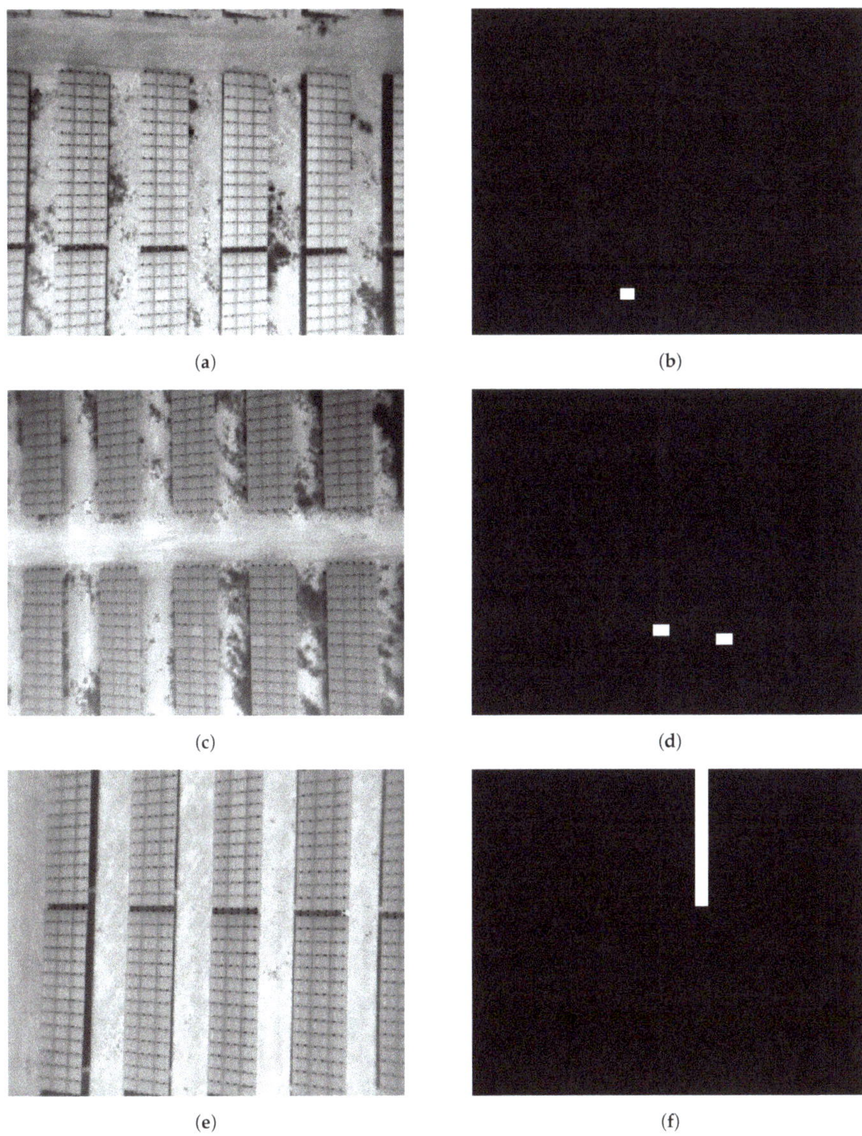

Figure 3. Examples of images from the dataset. Figures (**a,c,e**) are normalized thermal images. Figures (**a,c,e**) depict examples of masks, where the black color is the background that contains all the cells without anomalies and the white is the cells with anomalies. Figure (**b**) is an example of mask with a single anomaly cell; Figure (**d**) is a mask with two separated cells with anomalies; Figure (**f**) is a mask with continuous cells that present an anomaly.

The number of images per class are reported in Table 2.

Table 2. Anomalies statistics.

Class	Images
One Anomaly	841
Not Contiguous Cells with Anomalies	116
Contiguous Cells with Anomalies	52
Total Dataset	**1009**

3.2.2. Data Normalization

As already stated, the thermal data have several advantages compared to RGB data. However, a normalization and a transformation into a black and white image are required, for obtaining a single information channel. Table 3 shows the great variability of values within the thermal dataset: temperature values range between a minimum of 2.249 °C and a maximum of 103.335 °C, with a median equal to 44.21 °C. Figure 4 represents the histogram of temperatures of the whole dataset. Therefore, due to the great variability of values in the dataset, the thermal dataset was normalized in a range between 0 and 1, then transformed into grayscale images, i.e., with pixels having a value between 0 and 255. Examples of normalized thermal images are shown in Figure 3a,c,e.

Table 3. Thermal statistics of photovoltaic thermal image dataset.

Measure	Value (°C)
Minimum Temperature	2.249
Maximum Temperature	103.335
Average Temperature	43.860
Median	44.212
Standard Deviation	7.841

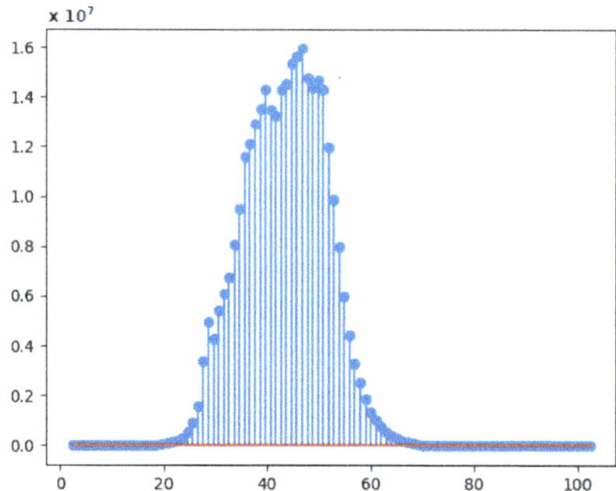

Figure 4. Histogram of the temperatures of the photovoltaic thermal image dataset. The X-axis shows the temperature values. The Y-axis reports the number of pixels of the dataset frames belonging to a temperature range. There are 100 intervals, each with a range of 1 degree, from the minimum to the maximum temperature values recorded in the dataset.

3.3. DNN-Based Solution for Anomaly Cells Detection

In this Subsection, we introduce the proposed deep learning-based solution for PV anomaly cells detection. In particular, the presence and the right position of an anomalous cell in a PV image is addressed as a segmentation task.

Image segmentation techniques take as input an image and output a mask with the predicted anomalous cells. Since it is a binary segmentation, the mask has pixels with values equal to 0 for the background and 1 for the anomalous cell. The DNNs specifically designed for image segmentation use convolutional neural networks for image classification as backbones for feature extraction, and on these backbones different kinds of feature combinations are constructed to achieve the segmentation result. CNNs are the most successful, well-known and widely used architectures in the deep learning domain, especially for computer vision tasks. They are a particular neural network that is able to extract discriminant features from data with convolution operations, so they can also be used as feature extraction networks. Usually a CNN is composed by three type of layers: convolutional layers, where a kernel of weights is convolved on inputs to extract discriminant features; non-linear layers, to learn the modeling of non-linear functions by the network; and finally, pooling layers, which reduce dimensions of a feature map by using statistical operations (mean, max). The units of every layer are locally connected, i.e., units receive weighted inputs from a small neighborhood (receptive field) of units of the previous layer. A CNN architecture is usually composed by stacking layers to form multi-resolution pyramids: the higher-level layers learn features from increasingly wider receptive fields. State-of-art CNN architectures are AlexNet, VGG , ResNet, MobileNet, and more recently, EfficientNet [31]. In this work, the backbone of the three segmentation networks is based on EfficientNet [31]. This network uses a mobile inverted bottleneck for the image classification task. Based on the backbone extracted features, the three segmentation methods that are compared for the development of our system are: UNet [15], LinkNet [17] and feature pyramid network (FPN) [32].

UNet is composed of a series of convolutional layers where the outputs of those layers are passed to a corresponding deconvolutional layer. In particular, a contracting path and expansive path are applied to generate a segmentation mask.

LinkNet was chosen because it is lightning-fast and is composed of a series of encoder and decoder blocks used to break down the image and build it back up before passing it through a few final convolutional layers. The structure of the network has been designed to minimize the number of parameters so that segmentation could be done in real time. Instead of a simple contracting path and expanding path, it is used a "link", which is inserted between the contracting paths and connects the result of the single step of contraction to the specular step of the expanding path.

Feature pyramid network (FPN) [32] is designed as creates a pyramid representation of the input image and on it apply the extraction network. It replaces the feature extractor of detectors like Faster R-CNN and generates multiple feature map layers (multi-scale feature maps) with better-quality information than the regular feature pyramid for object detection.

All these techniques give as output a single overall mask containing all the anomalous cells predicted in the same input image.

3.4. Mask Region-Based CNN for Anomaly Cells Detection

Instance segmentation has been chosen for this work. The main reason behind this solution is that the advantage for the operator is that it obtains the correct position of the anomalous cell. If compared to a common segmentation task, instance segmentation requires a mask to be created for each anomalous cell and within the same image. Conversely, image segmentation needs a further step to split all the defective cells calculated within the overall mask mentioned above. Following this assumption, Mask R-CNN was proven to be an effective and accurate network for solving these problems [13]. It is based on Faster R-CNN [33] and has an additional branch for predicting segmentation masks on each Region of Interest (RoI) in a pixel-to-pixel manner. This network generates three outputs: one for each candidate object, one for a class (considering both the label and a bounding-box offset) and one for the object mask. Additionally, it is comprises two parts: a region proposal network (RPN), which proposes a candidate object with a bounding box, and a binary mask classifier, which generates a mask for every class. Considering the specific case of anomaly cells detection, this network is not trained directly with image masks, but it needs the anomalous cell bounding boxes within the image: not a single mask, but a set of top-left and bottom-right coordinates of each bounding box. Furthermore, in order to be compared with image segmentation techniques, it also needs a post-processing step: all the anomalous predicted cells have to be merged into one overall mask.

3.5. Evaluation Metrics and Loss Function

The metrics taken into consideration vary according to the type of task to be solved and therefore the type of available output. For the image segmentation task the output is a total mask containing all the defective cells segmented for the same input image. In this case, pixel-based metrics used in state-of-the-art techniques are accuracy, precision, recall, and F1-score. However, these metrics can be misleading as the dataset is unbalanced, so the pixels belonging to damaged areas are far fewer than those concerning the non-damaged areas. To solve this problem we used other more suitable metrics: the Jaccard index (Equation (1)) and the Dice coefficient (Equation (2)).

The Jaccard index is a similarity measure on sets [34], and in the segmentation task the sets are the masks: the first one is that generated by the network and the second one is the ground truth mask.

$$J(A, B) = \frac{|A \cap B|}{|A \cup B|} \tag{1}$$

In Equation (1), A is the generated mask and B is the ground truth mask.

The Dice coefficient is a measure of the overlapping of two images, in this application the images are masks; the generated mask is A and the ground truth is B.

$$Dice = \frac{2|A \cap B|}{|A \cup B|} \tag{2}$$

These metrics are useful when, given an input image, the output is a single mask. This is not true for the Mask-RCNN, where a mask for each damaged identified cell is obtained. In this case, a post-processing phase is used to combine the masks into an overall mask and then calculate the metrics. The use of the Jaccard index and Dice coefficient, together with the publication of the thermal dataset, allows the scientific community to compare their approaches with the results of this work. For the training of the networks, starting from the metrics used to evaluate the performance, it is possible to use two cumulative loss functions, that is, a combination of the basic loss functions. The basic loss functions for the training of a network for image segmentation are: the Jaccard loss function (Equation (3)) and the Dice loss function (Equation (4)), described as follows:

$$JL(A, B) = 1 - J(A, B) \tag{3}$$

$$DL = 1 - Dice \tag{4}$$

In addition to these metrics, we used the Focal loss [35], suitable for segmentation tasks with unbalanced datasets where the background has a greater number of pixels than relative to the foreground. The Focal loss definition, in Equation (5), uses a posteriori probability p_t, which is the estimated probability for the class $y = 1$, where $y = \pm 1$. Focal loss uses an hyperparameter γ to tune the weight of different samples; the optimum value of γ, from [35], is 2.

$$FL(p_t) = -(1 - p_t)^\gamma log(p_t) \tag{5}$$

These basic losses are combined to obtain the two different loss functions used to train the networks. The first one is used to maximize the Dice and Jaccard coefficients and is detailed in Equation (6):

$$Loss_1 = \alpha * DL + (1 - \alpha) * JL \tag{6}$$

The second one is used to maximize the Dice coefficient and the focal loss over the different classes, and it is defined in Equation (7):

$$Loss_2 = DL + FL \tag{7}$$

4. Results and Discussion

In this Section, the results of the experiments conducted on the photovoltaic thermal images dataset are reported. In particular, two experiments were performed: the first one is based on the performance comparison of the three image segmentation networks (U-Net, LinkNet and FPN) and the second one involves the Mask R-CNN for the instance segmentation task. Finally, a comparative analysis of the networks is carried out. The photovoltaic thermal images dataset was split into three subset: 70% for training, 20% for validation and 10% for the final test. For both image and instance segmentation, the evaluation metrics used were the Dice and Jaccard indexes, as described in Section 3.5.

In the first experiment, the performances of three image segmentation networks were compared. These networks were implemented using TensorFlow and Keras and the training was carried out for 100 epochs, using $Loss_1$ (Equation (6)). The results achieved by these networks are summarized in Table 4 in terms of the Jaccard and Dice indexes. Results show that all networks showed good performance and are very similar: LinkNet slightly outperformed the others in terms of Jaccard index while U-Net was better than the others in terms of the Dice index.

Table 4. Evaluation of image segmentation on photovoltaic thermal images test set by using U-Net, LinkNet and FPN networks.

	Jaccard	Dice
U-Net	0.741	**0.841**
LinkNet	**0.748**	0.825
FPN	0.734	0.825

For the second experiment, we trained and tested the Mask R-CNN network. This network was also implemented in Keras and Tensorflow. In contrast to other DNNs, this is a network has been specifically developed for instance segmentation. For this reason, it is important to make a few remarks about these comparisons. First of all, the input of the network is the ground truth of the anomalous cells in form of bounding boxes, instead of the masks. Thus, starting from the masks it is necessary to have a preprocessing phase that allows to calculate the coordinates of these bounding boxes. The polygons obtained and their position in the reference image were finally saved in a json file. During the training, the batch size was fixed at 2 and the dataset was split as stated before. As described in Section 3.4, Mask R-CNN comprises several networks, and hence its loss function is defined as the sum of the losses of the different network components:

$$Loss_{total} = Loss_{cls} + Loss_{box} + Loss_{mask} \tag{8}$$

where $Loss_{cls}$ represents the loss of the classifier, $Loss_{box}$ is the loss of the regressor, and $Loss_{mask}$ is the loss of the segmentation branch.

The training was performed in three steps:

- Network trained from scratch;
- Network pretrained on the Microsoft Common Objects in the Context dataset (MS-COCO) [36], then retraining all layers;
- Network pretrained on the MS-COCO dataset, then retraining only the layers of the head section (the classifier section).

The technique of retraining a pre-trained network on another dataset is a transfer learning technique called fine tuning, and it is widely adopted in cases of small datasets. This technique generally allows to train a network faster than training from scratch. This approach proved to help in achieving excellent results in [37,38], using a Mask-RCNN pretrained on the MS-COCO dataset for their tasks. Another difference compared to other networks is that it obtains as output a mask for each anomalous predicted cell. For obtaining Jaccard and Dice metrics, a post-processing phase is needed to combine all the masks of the instances into a single overall mask. Table 5 reports the Instance segmentation results on the photovoltaic thermal images test set by using a Mask-RCNN network. The results of the three training approaches are reported, in terms of Jaccard and Dice metrics.

Table 5. Instance segmentation results on the photovoltaic thermal images test set by using a Mask-RCNN network.

	Jaccard	Dice
Mask-RCNN (all)	0.382	0.493
Mask-RCNN (pretrain-all)	0.106	0.175
Mask-RCNN (pretrain-head)	**0.499**	**0.605**

The results show that using a pre-trained network and re-training only the head part allows obtaining a good instance segmentation network: it achieved 0.499 on the Jaccard index and 0.605 on the Dice index. These performances are higher than the network trained from scratch. The tests also show that totally re-training a pre-trained network could lead to worse results than training it from scratch.

Finally, Table 6 presents a comparative analysis of the performance of the best networks for both segmentation approaches. The trainable parameters and the training time are also reported. For this comparison, the network chosen is UNet, for its Jaccard and Dice index metrics, and Mask RCNN pre-trained on the MS-COCO dataset and re-trained only on the the head part. The results reveal that the U-Net outperformed the other approaches. However, Mask-RCNN has the key advantage that it directly outputs the position of each single predicted cell. Conversely, UNet outputs a single overall mask, but through a post-processing step based on image processing techniques, it can be easily split into the individual predicted cells.

Table 6. Comparative analysis of the performance of the best networks for both segmentation approaches.

	Jaccard	Dice	Trainable Params	Total Params	Training Time	Test Time
U-Net]	**0.741**	**0.841**	74.7 Million	75 Million	192 s/epoch	15 ms
Mask-RCNN	0.499	0.605	21 Million	44.6 Million	212 s/epoch	24 ms

Figure 5 depicts the training trend of the UNet network, concerning the loss function, the Jaccard index and the Dice index. It can be noticed that after only 30 epochs the network tends to the convergence.

Figure 6 allows a visual analysis of the results obtained by the UNet network on the test set. It represents some examples of test images, their ground truth and the relative predicted mask. It is possible to deduce that for a human operator it is easy to understand its exact position within the plan (Figure 6a,b). This leads to many advantages in terms of time and efforts. Figure 6c shows that the network may have false positives in the predicted mask, i.e., some areas are miss-classified as anomalous cells. These false positives usually have a very small area.

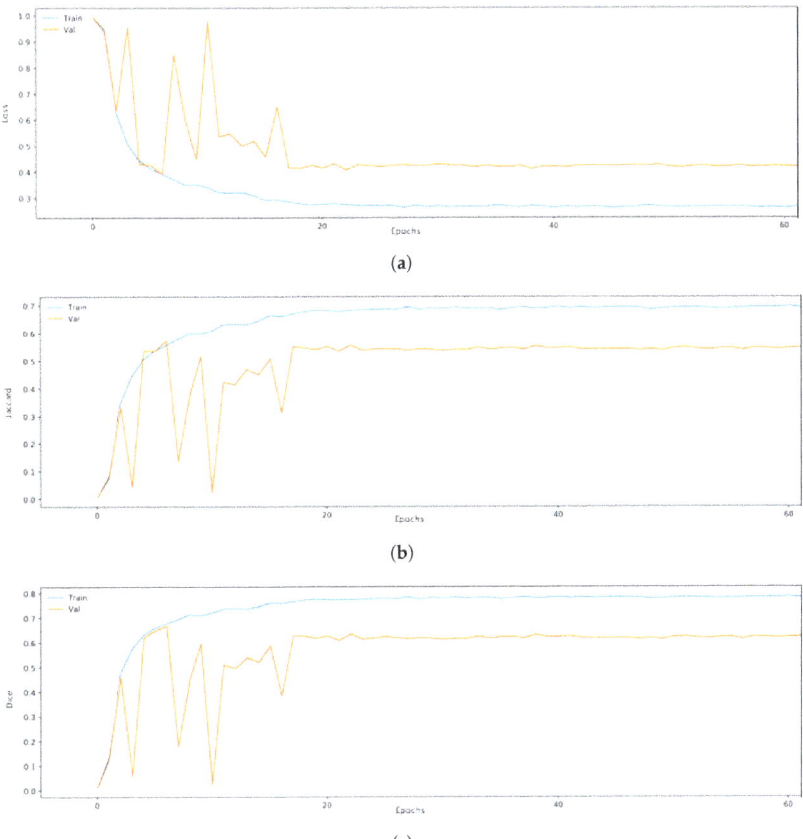

(a)

(b)

(c)

Figure 5. Training of the U-Net. The figure shows the trend of the training concerning the loss function (**a**), the Jaccard index (**b**) and the Dice index (**c**).

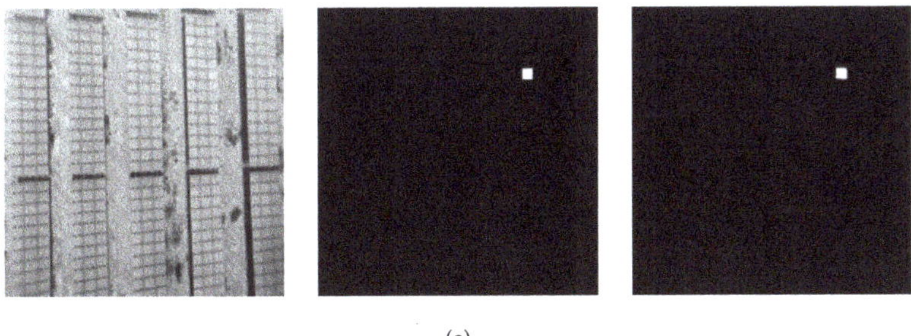

(a)

Figure 6. *Cont.*

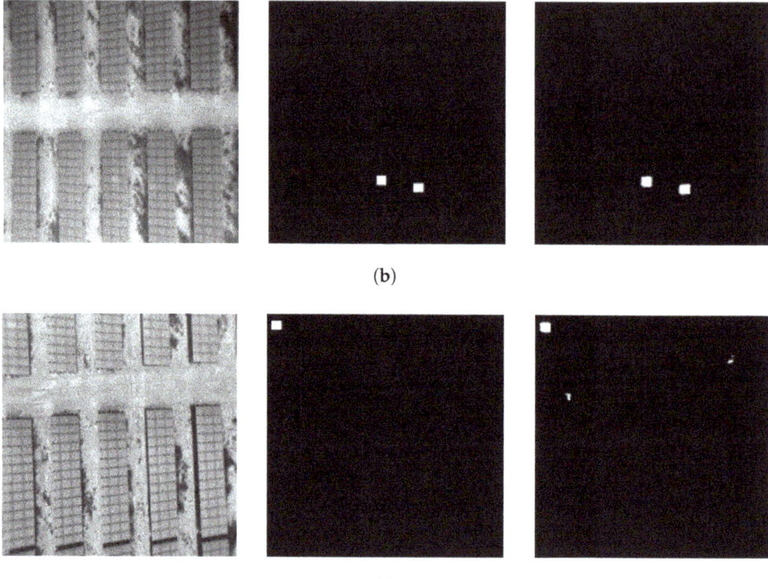

(b)

(c)

Figure 6. UNet performance on test set images, with ground truth and predicted mask. The masks of (**a**,**b**) have been correctly predicted. (**c**) depicts some missclassified areas.

5. Conclusions and Future Works

In this study, solAIr, an artificial intelligence UAV-based inspection system was presented, which is capable of detecting faults in large-scale PV plants. To achieve such results, a DNN deep-learning based module was developed, and was designed to exert instance segmentation. The proposed solution was properly evaluated against existing solutions through a comparative study. The experimental results confirm its effectiveness and suitability to the diagnosing of thermal images of PV plants. In particular, the networks chosen obtained high values on the Jaccard and Dice indices. The proposed approach for defect analysis can be an essential aid to assist operators for O&M operations, reducing cost and errors arising from manual operations. Considering that, nowadays, inspections are entrusted to visual inspections, our approach will both reduce the overall costs of PV module maintenance and increase the efficiency of PV plants. Considering that instance segmentation through deep learning has never been applied in this field before, this study advances the body of knowledge and opens up promising scenarios for the management of clear energies. The work also presents some drawbacks: first of all, we only deal with binary segmentation: a pixel can be classified either as a damaged cell or as a background. Notwithstanding, this issue can be easily overcome. In fact, this framework is already prepared for a future multi-class segmentation: for example, detecting different types of cell anomalies, as described in the Section 2. A further consideration that can be made concerning the used dataset: probably creating a mask that combines several defective cells (Figure 3f) could add an error in the training of the network, because the pixels of conjunction between the cells should not be part of this mask. Hence the performance of the network will improve as soon as the masks of this type of defect are improved. The output results of the proposed experiments can be easily integrated within a dedicated geographical information system (GIS) specifically designed for operation and maintenance (O&M) activities in PV Plants. Indeed, having the geolocation for each image, the management of the detected faulty cells can be facilitated. Moreover, thermal data, which are processed with their raw values, still require a processing phase in the office. Given the good computational performances described in Table 6, an on-board integration in the UAV platform can be foreseen for on-site inspection operation, with minimized implementation hurdles. Additionally, the robustness and the reliability of the proposed UAV-based inspection system, along with the deep-learning anomaly cells detection solution, needs to be further validated and improved through extensive field assessments. A further improvement can be made by exploiting the data analysis of the real-time electrical measurements of operating PV modules, obtained from the underlying system monitoring infrastructure. Such data can be used in conjunction with the proposed solution to improve the performances of the fault detection system.

Author Contributions: Conceptualization, R.P. and M.P.; methodology, R.P. and M.P.; software, A.F.; validation, P.Z.; data curation, F.P.; writing—original draft preparation, R.P. and M.P.; visualization, A.F.; supervision, P.Z. All authors have read and agreed to the published version of the manuscript.

Funding: This research received no external funding.

Acknowledgments: This work was supported by Fly Engineering s.r.l. (www.flyengineering.it).

Conflicts of Interest: The authors declare no conflict of interest.

References

1. Li, X.; Yang, Q.; Lou, Z.; Yan, W. Deep Learning Based Module Defect Analysis for Large-Scale Photovoltaic Farms. *IEEE Trans. Energy Convers.* **2018**, *34*, 520–529. [CrossRef]
2. Wang, M.; Cui, Q.; Sun, Y.; Wang, Q. Photovoltaic panel extraction from very high-resolution aerial imagery using region–line primitive association analysis and template matching. *ISPRS J. Photogramm. Remote Sens.* **2018**, *141*, 100–111. [CrossRef]
3. Tsanakas, J.A.; Ha, L.; Buerhop, C. Faults and infrared thermographic diagnosis in operating c-Si photovoltaic modules: A review of research and future challenges. *Renew. Sustain. Energy Rev.* **2016**, *62*, 695–709. [CrossRef]
4. Sharma, V.; Chandel, S. Performance and degradation analysis for long term reliability of solar photovoltaic systems: A review. *Renew. Sustain. Energy Rev.* **2013**, *27*, 753–767. [CrossRef]
5. Tsanakas, J.A.; Vannier, G.; Plissonnier, A.; Ha, D.L.; Barruel, F. Fault diagnosis and classification of large-scale photovoltaic plants through aerial orthophoto thermal mapping. In Proceedings of the 31st European Photovoltaic Solar Energy Conference and Exhibition 2015, Hamburg, Germany, 14–18 September 2015; pp. 1783–1788.
6. Köntges, M.; Kurtz, S.; Packard, C.; Jahn, U.; Berger, K.A.; Kato, K.; Friesen, T.; Liu, H.; Van Iseghem, M.; Wohlgemuth, J.; et al. *Review of Failures of Photovoltaic Modules*; International Energy Agency: Paris, France, 2014.
7. Grimaccia, F.; Leva, S.; Dolara, A.; Aghaei, M. Survey on PV modules' common faults after an O&M flight extensive campaign over different plants in Italy. *IEEE J. Photovoltaics* **2017**, *7*, 810–816.
8. Grimaccia, F.; Leva, S.; Niccolai, A.; Cantoro, G. Assessment of PV plant monitoring system by means of unmanned aerial vehicles. In Proceedings of the 2018 IEEE International Conference on Environment and Electrical Engineering and 2018 IEEE Industrial and Commercial Power Systems Europe (EEEIC/I&CPS Europe), Palermo, Italy, 12–15 June 2018; pp. 1–6.
9. Piccinini, F.; Pierdicca, R.; Malinverni, E.S. A Relational Conceptual Model in GIS for the Management of Photovoltaic Systems. *Energies* **2020**, *13*, 2860. [CrossRef]
10. Liciotti, D.; Paolanti, M.; Pietrini, R.; Frontoni, E.; Zingaretti, P. Convolutional networks for semantic heads segmentation using top-view depth data in crowded environment. In Proceedings of the 2018 24th International Conference on Pattern Recognition (ICPR), Beijing, China, 20–24 August 2018; pp. 1384–1389.
11. Paolanti, M.; Romeo, L.; Martini, M.; Mancini, A.; Frontoni, E.; Zingaretti, P. Robotic retail surveying by deep learning visual and textual data. *Robot. Auton. Syst.* **2019**, *118*, 179–188. [CrossRef]
12. Lyu, Y.; Vosselman, G.; Xia, G.S.; Yilmaz, A.; Yang, M.Y. UAVid: A semantic segmentation dataset for UAV imagery. *ISPRS J. Photogramm. Remote Sens.* **2020**, *165*, 108–119. [CrossRef]
13. He, K.; Gkioxari, G.; Dollár, P.; Girshick, R. Mask r-cnn. In Proceedings of the IEEE International Conference on Computer Vision, Venice, Italy, 22–29 October 2017; pp. 2961–2969.

14. Pierdicca, R.; Malinverni, E.; Piccinini, F.; Paolanti, M.; Felicetti, A.; Zingaretti, P. Deep Convolutional neural network for automatic detection of damaged photovoltaic cells. In Proceedings of the International Archives of the Photogrammetry, Remote Sensing and Spatial Information Sciences, 2018 ISPRS TC II Mid-term Symposium "Towards Photogrammetry 2020", Riva del Garda, Italy, 4–7 June 2018; Volume 42.

15. Ronneberger, O.; Fischer, P.; Brox, T. U-net: Convolutional networks for biomedical image segmentation. In Proceedings of the International Conference on Medical image computing and computer-assisted intervention, Munich, Germany, 5–9 October 2015; pp. 234–241.

16. Yang, Y.; Wang, C.; Gong, L.; Zhou, X. FPNet: Customized Convolutional Neural Network for FPGA Platforms. In Proceedings of the 2019 International Conference on Field-Programmable Technology (ICFPT), Tianjin, China, 9–13 December 2019; pp. 399–402.

17. Chaurasia, A.; Culurciello, E. Linknet: Exploiting encoder representations for efficient semantic segmentation. In Proceedings of the 2017 IEEE Visual Communications and Image Processing (VCIP), St. Petersburg, FL, USA, 10–13 December 2017; pp. 1–4.

18. Rogotis, S.; Ioannidis, D.; Tsolakis, A.; Tzovaras, D.; Likothanassis, S. Early defect diagnosis in installed PV modules exploiting spatio-temporal information from thermal images. In Proceedings of the 12th Quantitative InfraRed Thermography Conference, Bordeaux, France, 7–11 July 2014; Volume 7, p. 14.

19. Tsanakas, J.A.; Chrysostomou, D.; Botsaris, P.N.; Gasteratos, A. Fault diagnosis of photovoltaic modules through image processing and Canny edge detection on field thermographic measurements. *Int. J. Sustain. Energy* **2015**, *34*, 351–372. [CrossRef]

20. Kim, D.; Youn, J.; Kim, C. Automatic fault recognition of photovoltaic modules based on statistical analysis of UAV thermography. In Proceedings of the International Archives of the Photogrammetry, Remote Sensing and Spatial Information Sciences 2017, Tehran, Iran, 7–10 October 2017; Volume 42, p. 179.

21. Leotta, G.; Pugliatti, P.; Di Stefano, A.; Aleo, F.; Bizzarri, F. Post Processing Technique for Thermo-Graphic Images Provided by Dron Inspections. In Proceedings of the 31st European Photovoltaic Solar Energy Conference and Exhibition, Hamburg, Germany, 14–18 September 2015; pp. 1799–1803.

22. Zhao, J.; Ye, F. Where ThermoMesh meets ThermoNet: A machine learning based sensor for heat source localization and peak temperature estimation. *Sens. Actuators A Phys.* **2019**, *292*, 30–38. [CrossRef]

23. Gao, X.; Munson, E.; Abousleman, G.P.; Si, J. Automatic solar panel recognition and defect detection using infrared imaging. In Proceedings of the Automatic Target Recognition XXV. International Society for Optics and Photonics, Baltimore, MD, USA, 20–22 April 2015; Volume 9476, p. 94760O.

24. Buerhop, C.; Pickel, T.; Dalsass, M.; Scheuerpflug, H.; Camus, C.; Brabec, C.J. aIR-PV-check: A quality inspection of PV-power plants without operation interruption. In Proceedings of the 2016 IEEE 43rd Photovoltaic Specialists Conference (PVSC), Portland, OR, USA, 5–10 June 2016; pp. 1677–1681.

25. De Oliveira, A.K.V.; Aghaei, M.; Rüther, R. Automatic Fault Detection of Photovoltaic Array by Convolutional Neural Networks During Aerial Infrared Thermography. In Proceedings of the 36th European Photovoltaic Solar Energy Conference and Exhibition, Marseille, France, 9–13 September 2019.

26. Herraiz, Á.H.; Marugán, A.P.; Márquez, F.P.G. Optimal Productivity in Solar Power Plants Based on Machine Learning and Engineering Management. In Proceedings of the International Conference on Management Science and Engineering Management, Melbourne, VIC, Australia, 1–4 August 2018; pp. 983–994.

27. Wei, S.; Li, X.; Ding, S.; Yang, Q.; Yan, W. Hotspots Infrared detection of photovoltaic modules based on Hough line transformation and Faster-RCNN approach. In Proceedings of the 2019 6th International Conference on Control, Decision and Information Technologies (CoDIT), Paris, France, 23–26 April 2019; pp. 1266–1271.

28. Dunderdale, C.; Brettenny, W.; Clohessy, C.; van Dyk, E.E. Photovoltaic defect classification through thermal infrared imaging using a machine learning approach. In *Progress in Photovoltaics: Research and Applications*; Wiley Online Library: Hoboken, NJ, USA, 2019.

29. Mehta, S.; Azad, A.P.; Chemmengath, S.A.; Raykar, V.; Kalyanaraman, S. Deepsolareye: Power loss prediction and weakly supervised soiling localization via fully convolutional networks for solar panels. In Proceedings of the 2018 IEEE Winter Conference on Applications of Computer Vision (WACV), Lake Tahoe, NV, USA, 12–15 March 2018; pp. 333–342.

30. Buerhop-Lutz, C.; Deitsch, S.; Maier, A.; Gallwitz, F.; Brabec, C. A benchmark for visual identification of defective solar cells in electroluminescence imagery. In Proceedings of the 35th European PV Solar Energy Conference and Exhibition, Brussels, Belgium, 24–27 September 2018; Volume 12871289.

31. Tan, M.; Le, Q.V. Efficientnet: Rethinking model scaling for convolutional neural networks. *arXiv* **2019**, arXiv:1905.11946.

32. Lin, T.Y.; Dollár, P.; Girshick, R.; He, K.; Hariharan, B.; Belongie, S. Feature pyramid networks for object detection. In Proceedings of the IEEE conference on computer vision and pattern recognition, Honolulu, HI, USA, 21–26 July 2017; pp. 2117–2125.

33. Ren, S.; He, K.; Girshick, R.; Sun, J. Faster r-cnn: Towards real-time object detection with region proposal networks. In Proceedings of the Advances in Neural Information Processing Systems, Montreal, QC, Canada, 7–12 December 2015; pp. 91–99.

34. Jaccard, P. Étude comparative de la distribution florale dans une portion des Alpes et des Jura. *Bull Soc. Vaudoise Sci. Nat.* **1901**, *37*, 547–579.

35. Lin, T.Y.; Goyal, P.; Girshick, R.; He, K.; Dollár, P. Focal loss for dense object detection. In Proceedings of the IEEE international conference on computer vision, Venice, Italy, 22–29 October 2017; pp. 2980–2988.

36. Lin, T.Y.; Maire, M.; Belongie, S.; Hays, J.; Perona, P.; Ramanan, D.; Dollár, P.; Zitnick, C.L. Microsoft coco: Common objects in context. In Proceedings of the European Conference on Computer Vision, Zurich, Switzerland, 6–12 September 2014; pp. 740–755.

37. Zhang, W.; Witharana, C.; Liljedahl, A.; Kanevskiy, M. Deep convolutional neural networks for automated characterization of arctic ice-wedge polygons in very high spatial resolution aerial imagery. *Remote Sens.* **2018**, *10*, 1487. [CrossRef]

38. Jaiswal, A.; Tiwari, P.; Kumar, S.; Gupta, D.; Khanna, A.; Rodrigues, J. Identifying pneumonia in chest X-rays: A deep learning approach. *Meas. J. Int. Meas. Confed.* **2019**, *145*, 511–518. [CrossRef]

Publisher's Note: MDPI stays neutral with regard to jurisdictional claims in published maps and institutional affiliations.

Article

An Interval-Arithmetic-Based Approach to the Parametric Identification of the Single-Diode Model of Photovoltaic Generators

Martha Lucia Orozco-Gutierrez

Escuela de Ingeniería Eléctrica y Electrónica, Universidad del Valle, Cali 760036, Colombia; martha.orozco@correounivalle.edu.co; Tel.: +57-2-321-21-00

Received: 21 January 2020; Accepted: 13 February 2020; Published: 19 February 2020

Abstract: Parametric identification of the single diode model of a photovoltaic generator is a key element in simulation and diagnosis. Parameters' values are often determined by using experimental data the modules manufacturers provide in the data sheets. In outdoor applications, the parametric identification is instead performed by starting from the current vs. voltage curve acquired in non-standard operating conditions. This paper refers to this latter case and introduces an approach based on the use of interval arithmetic. Photovoltaic generators based on crystalline silicon cells are considered: they are modeled by using the single diode model, and a divide-and-conquer algorithm is used to contract the initial search space up to a small hyper-rectangle including the identified set of parameters. The proposed approach is validated by using experimental data measured in outdoor conditions. The information provided by the approach, in terms of parametric sensitivity and of correlation between current variations and drifts of the parameters values, is discussed. The results are analyzed in view of the on-site application of the proposed approach for diagnostic purposes.

Keywords: parametric identification; single-diode model; interval arithmetic; photovoltaic systems

---

1. Introduction

Photovoltaic (PV) array modeling is crucial in many fields, including the prediction of energy production [1], the design, the control [2] and the diagnosis [3]. The increase of the PV cells would be desirable, in a context where many types of technologies have been developing, although 90%–95% of the market is still dominated by mono-crystalline and poly-crystalline silicon technologies [4]. The mono-crystalline PV commercial modules reach efficiencies between 15% and 22%; meanwhile, poly-crystalline technology goes up to the efficiency range of 14%–20%. The economies of scale of its main material, silicon, make crystalline silicon cells more affordable and highly efficient compared to other materials [5]. Other technologies derived from crystalline silicon technologies have been gaining importance in the research and commercial fields, such as half-cell, double glass and bifacial [4]. On the other hand, thin film technologies, e.g., amorphous silicon, CdS/CdTe and CIS, represent close to 5%–10% of the market [4]. Emerging technologies, e.g., organic and perovskite ones, offer interesting perspectives in terms of efficiency [6], but some barriers still need to be overcome, especially durability and price [5]. The approach proposed in this paper refers to PV generators based on crystalline silicon cells, which represent the largest part of the market. Different models have been studied in the scientific publications to represent PV modules based on crystalline silicon cells. The single diode model (SDM) offers a reasonable trade-off between accuracy and degree of non linearity, such that it is widely used in literature. It involves five parameters, which are related to the photo-induced current, the P-N junction and the losses. These parameters are in turn dependent on other ones related to the cell material and the environmental conditions—the irradiance and the temperature; see [7,8]. The double diode model (DDM) allows one to model the dark current losses and the effect of pair generation—recombination in the space charge region [9], but at the cost of an increase in the number of parameters, increasing from five for the SDM, to seven. A more complicated model can be used in a case where the PV cells' behavior at negative voltage values has to be accounted for. A further generator is included in this model [10], so that the parameters required become eight.

In this paper, the SDM is preferred to the DDM because of the features mentioned above and also because the PV array working conditions considered are uniform; thus, the model proposed in [10] becomes superfluous. A key operation for an accurate SDM-based simulation of the PV array is the identification of the five model parameters. This is very often done by employing data that are provided by the PV module manufacturer through the data sheet. These experimental measurements refer to a specific operating conditions of the cells, called standard test conditions (STC).

In the literature, parametric identification is done by using analytical methods or fitting techniques [11]. In Table 1 a comparison among such approaches is given by referring to the implementation complexity, the convergence speed, the robustness when noisy data are considered, the impact of the initial conditions and the requirements of algorithmic setting. Analytical approaches are based on a set of simplified equations leading to explicit formulas allowing one to calculate the five parameters of the SDM without using any iterative method [12]. Some approaches consider the SDM lossless model, or scale down the order of the SDM by considering an infinite value of the shunt resistance or by neglecting the series resistance [13]. The most common simplification consists of supposing that the short-circuit current (I_{sc}) value is equal to the photo-induced current (I_{ph}) [14]. A set of equations is derived at the main points of the current vs. voltage (I-V) curve: at the maximum power point (MPP), at the short circuit operating point through I_{sc} and in open circuit conditions through V_{oc}. Noisy I-V data may have a significant effect on parameters values. This is the case, for instance, for a series and for the shunt resistances whose values are related to the slopes of the I-V curve in an open circuit and in short circuit conditions, respectively. Additionally, the so called *translation equations* have been considered in some papers to relate the I-V curve in non-standard conditions to the five SDM parameters that are scaled according to the irradiance and temperature conditions the I-V curve refers to [14–18]. Simplified and direct equations make the implementation of the method suitable, even for an embedded processor, at the cost of a reduced accuracy.

Many other approaches to the SDM parametric identification are based on optimization methods, which are usually aimed at the root mean square error (RMSE) between the simulated I-V curve and the experimental curve minimization. The convergence of the algorithm depends on factors such as the guess condition, the objective function and the algorithm itself. Three main approaches are proposed: the one using non-linear minimization algorithms [19,20], the one using heuristics approaches [21–24] and the adoption of hybrid methods [25]. The non-linear algorithms are computationally expensive [19,20], but that allows for solving numerically, the set of non-linear equations of SDM. In [20], *Matlab* embedded Levenberg–Marquardt and Gauss-Newton nonlinear equation solvers were used to manage the SDM equations. The parameters in STC were obtained from modules' data sheets, and *translation equations* [15,16] were used to obtain the SDM model in other operating conditions. Noisy data affect the confidence intervals of the solutions achieved by these algorithms. The termination conditions and the related parameters have to be chosen in order to have a trade-off between computation time and accuracy. Low values of the convergence thresholds and high values assigned to the maximum number of iterations are compatible with off-line identification purposes. Moreover, the iterative methods, such as Newton methods, are less complex, but they might be trapped in local optima and show a high dependency from the guess solution used. For instance, in [26] R_s is neglected, so that four equations to determine four SDM parameters are proposed, the series resistance value being fitted by calculating the power error between the SDM and the experimental measurements in a iterative way. Nonlinear minimization algorithms are often used to fit I-V experimental curves, the objective function to minimize being the error between the model and experimental data. Trust-region and Levenberg-Marquardt methods are widely used, but they require a good guess solution. Simulation platforms, e.g., *Matlab* and *Mathematica*, provide curve fitting tools to perform offline parametric identification. Recently, soft computing methods, such as artificial neural networks [21,22], genetic algorithms [23] and particle swarm optimization (PSO) [23], among others, have been employed more and more frequently. Such approaches are not suitable for online operation because of the computational complexity of the stochastic algorithms. The approaches introduced in [21,22] operate on suitable sets of I-V curves, related to a specific module's operation, to train neural networks. To determine the amount of training data and the numbers of layers and neurons is a challenging task. In genetic algorithms [23], fixing selection, reproduction and mutation operators and values of the related parameters is challenging as well. Settings such as population size, iteration number and mutation rate, among others, have to be well adjusted to prevent the algorithm from stalling. Therefore, in terms of setting of the heuristic algorithms, several initial guess values have to be designed by an expert and/or through a trial and error procedure. By combining different techniques, some weaknesses are reduced. For example, in [25] the global exploration capabilities of the soft computing algorithm artificial bee colony (ABC) allowed it to reduce the space for exploring solutions, and local searching was done by the trust-region reflective algorithm, thereby improving accuracy, convergence and reliability. Unfortunately, there is not a consensus about the improvement of the computation time achieved by hybrid approaches.

The difficulty of the parametric identification comes from the high non linearity of the SDM and from the fact that the values of the parameters have very different orders of magnitude. With respect to the identification performed on the basis of the STC experimental data that are usually available in PV modules data sheets, the parametric identification using data acquired while the PV module is working in outdoor conditions show different features. Indeed, the whole I-V curve is usually available and irradiance (G) and temperature (T) values at which the PV module is working might be also given.

Interval arithmetic (IA) is a mathematical approach that is used in many contexts for evaluating the propagation of the uncertainty affecting input data on the output of a given system. Moreover, it has been used for tolerance analysis and design in the context of electrical and electronic engineering; e.g., in [27–29]. By IA, parameters assume values that are not real numbers, but intervals limited by a lower and an upper bound: in an interval the parameter may assume any value with the same probability. In [28] an evolutionary approach to worst case tolerance design of magnetic devices is presented. The algorithm improves on the classical nominal design, accounting for parameter variations and tolerances, so that the system performance does not exceed upper and lower specifications imposed in advance by the designer. In [27], IA is used to perform tolerance analysis and design and to evaluate the production yield. In [29], an IA based estimation state in power distribution networks with high penetration of photovoltaic generators is proposed. In this case, IA is adopted to deal with measurement uncertainty. The proposed method allows one to determine the upper and lower bounds of state variables, which is helpful for providing operators the confidence that the actual value variable is not exceeding the voltage security constraint, thereby improving the network operation for the case of uncertain inputs.

IA-based parametric identification has never been used in the outdoor PV context, but it can be helpful for designing an algorithm profiting from IA features, thereby giving a reliable result with little computational effort. The IA based approach presented in this paper starts from a large volume in the parameter search space and contracts it by means of a divide-and-conquer (*D&C*) strategy up to converge to a tight hyper-rectangle including the experimental measurements in the I-V plane. The proposed IA based *D&C* algorithm requires the user to fix the initial intervals for the five SDM parameters and two thresholds for the feasibility and the termination conditions respectively. The initial intervals, which define the search parameters' space, are contracted towards the identified set, if they are included in the search space. Otherwise, the IA based *D&C* algorithm informs the user about the guaranteed infeasibility of the whole search space. To fix the search space is obviously easier for a not-so-skilled user than to provide a guess solution that is quite close to the final one, as is required by gradient-based minimization approaches. This feature is very helpful, especially for some parameters, e.g., saturation current and thermal voltage, that greatly depend on cells material and on operating conditions. As for the feasibility condition, the desired amount of experimental data contained in the IA computed I-V boundaries depends on the application, and it can be fixed as greater than 85%. In the same way, the threshold to fix in the termination condition is chosen as the desired resolution of the interval solution. *D&C* strategy allows one to evaluate solutions separately, so parallel computing is suitable to decrease the computation time without sacrificing the size of search space. This is a distinctive feature compared to the other approaches, which in general have to make a compromise between accuracy and computation time.

Table 1. Comparison among different approaches to parametric identification of SDM.

Methods/ Features	Core of the Procedure	Implemen- Tation Complexity	Speed Conver- Gence	Robustness with Noisy Data	Initial Condition Impact	Require- Ment of Algorithm Setting
Analytical	Set of equations solved explicitly	Low	High	Medium	High	Medium
Iterative non-linear minimization	Equations must be solved by numerical methods	High	Low	Medium	High	Low
Heuristics	Fitting I-V curve model to measured data by a soft computing algorithm	High	Medium	Medium	High	High
Hybrids	Combining non-linear minimization and heuristics	High	Medium/ Low	Medium	High	High
Proposed IA Based D&C	Parameters are intervals divided and tested using feasibility conditions	Low	Medium/ High	High	Low	Low

The paper is organized as follows: In the first section an introduction of SDM is done. Then, IA theory is briefly recalled and it is applied to the SDM. Later on, *D&C* algorithm is presented. In Section 4, the proposed method using IA and *D&C* algorithm is detailed. In Section 5, the results obtained to estimate R_s, R_h, I_{sat}, B and I_{ph} parameters of the SDM model are analyzed. The sixth section proposes a discussion about the results presented in the paper and closes with the conclusions.

2. Photovoltaic Generator Single Diode Model

Figure 1 shows the SDM circuit: it includes the photoinduced current generator I_{ph}, which models the photovoltaic effect; a diode D modeling the P-N junction; and the resistances R_s and R_h representing the ohmic losses and the recombination losses respectively. Thus, the following five parameters appear in the model:

$I_{sat,d}$: saturation current in the P-N junction;

I_{ph}: photo-induced current;

R_s: series resistance;

R_h: parallel resistance;

B: it includes the ideality factor n, which is the fifth parameter to be identified. It is: $B = N_s \cdot n \cdot k \cdot T/q$, where N_s is the number of series connected cells, T is the cells operating temperature, k is the Boltzmann constant and q is the electron charge.

It is worth noting that the five parameters mentioned above show some dependencies from physical parameters that are typical of the semiconducting material used for the cells' fabrication and also from irradiance G and temperature T. As in the majority of the literature concerning I-V curve based parametric identification, in this paper also, the identification focuses on the five parameters mentioned above, thereby neglecting their dependencies on other physical parameters. This further correlation, and the dependency on G and T especially, can be exploited after having identified the set $\{I_{sat,d}, I_{ph}, R_s, R_h, B\}$ to the aim of having, in turn, the values of the physical parameters, including G and T.

The output current I of the PV array is obtained by combining the Kirchhoff voltage and current laws and the characteristic equations of the components appearing in the SDM.

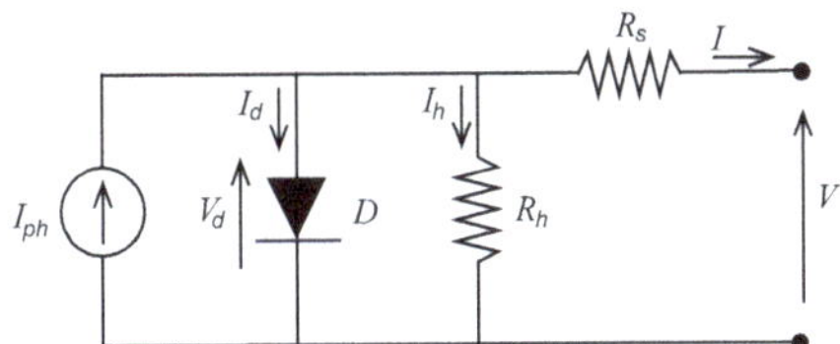

Figure 1. Circuit model of a PV module based on single-diode.

$$I = I_{ph} - I_d - I_h \tag{1}$$

$$I_d = I_{sat,d}(e^{V_d/B} - 1) \tag{2}$$

$$V_d = V + R_s \cdot I \tag{3}$$

$$I_h = \frac{V_d}{R_h} = \frac{V + R_s \cdot I}{R_h} \tag{4}$$

In [30] it is shown that the resulting function expressing the relationship between the current I and the voltage V at the PV generator terminals is implicit, but the Lambert W-function is useful for achieving an explicit non-linear relation between I and V, which is given in (5).

$$I = \frac{R_h(I_{ph} + I_{sat,d}) - V}{R_h + R_s} - \frac{B}{R_s} LambertW(\theta) \tag{5}$$

wherein: $\theta = \{(R_s//R_h)I_{sat,d}e^{[(R_hR_s(I_{ph}+I_{sat,d})+R_hV)/B(R_h+R_s)]}\}/B$

Later on, without loss of generality, the discussion is referred to one PV module. The set of unknown five parameters is $P = \{I_{sat,d}, I_{ph}, R_s, R_h, B\}$.

In Figure 2 a PV module I-V curve is shown by blue marks: it has been obtained by placing the values listed in Table 2 into Equation (5). The parameters in Table 2 refer to a 140 W PV Yingli solar panel working in STC, which have been obtained by the method proposed in [31] in STC. This PV module consists of 36 polycrystalline solar cells connected in series. In the same figure, the I-V curves corresponding to 30% variations of the parameters R_s, R_h, I_{sat} and B are also shown in magenta, cyan, red and black respectively. The I-V curve exhibits a significant sensibility with respect to variations of B and I_{sat} in proximity to the MPP and a dependency on R_s in the high voltage range. The parameter I_{ph} depends on, almost directly, irradiance, and its value is usually assumed to be equal to the short-circuit current [2].

In the literature, SDM parametric identification of the parameters has been often addressed by minimization algorithms, which are aimed at fitting the experimental I-V curve with the one generated by the SDM. The result is a set of five real values, one for each of the five parameters in the SDM (Table 2). IA, instead, should be used to identify the parameters by starting from the I-V curve, and by exploiting the IA properties, guaranteeing that the I-V ranges correspond to the set of parameters bound to the experimental measurements. Later on, the main IA features and properties are recalled in order to appreciate how they are suitably exploited in the PV parametric identification context.

Table 2. STC Parameters for a 140 W PV Yingli solar panel.

Parameter	Meaning	Value
$I_{sat,d}[A]$	Saturation current in cell	5.359647×10^{-10}
$I_{ph}[A]$	Photoinduced current	8.3
$R_s[\Omega]$	Series resistance	0.1793
$R_h[\Omega]$	Shunt resistance	150.70
B	$N_s \cdot n \cdot k \cdot T / q$	1.0532

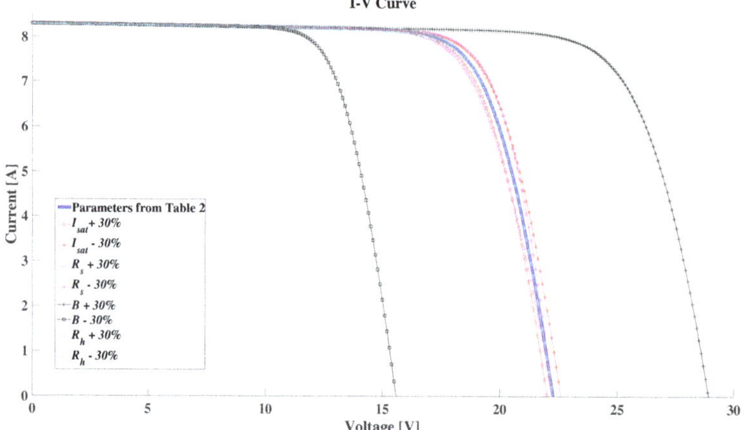

Figure 2. I-V curves of a PV module simulated using the parameters of Table 2 with 30% variations.

3. Interval Arithmetic for I-V Curve Representation

The basic mathematical entity used in IA is the interval. Thus, the parameters appearing in the model can be treated as intervals $[X]$ instead of real numbers X. The approach consists of treating parameters or variables as having ranges of values, instead of discrete values. The bounds of the interval $[X]$, using the nomenclature proposed in the IEEE Std 1788.1 [32], are called $\underline{x}$ and $\overline{x}$. Thus, $[X] = [\underline{x}, \overline{x}]$; it is defined by $x = \{x \in \Re / \underline{x} \leq x \leq \overline{x}\}$. Basic arithmetic operations among interval variables are well defined in the IA foundations [33]. The operation among two intervals results in an interval too, having the property that it contains all the possible results obtained by the combination of all the values included in the intervals corresponding to the operands. The values in the intervals are assumed to have the same probability of occurring, so that the probabilities of the values are uniform. This means that all the values in $[X]$ are equi-probable. IA theory also shows that the simple representation of the operands, which consists of the lower and of the upper bound of the intervals thereof, does not allow one to take into account any correlation among the variables. As a consequence of this, the IA result is an over-estimation of the true range of the result. This means that the IA result is guaranteed to contain all the possible results of the operation, but the over-estimation might be too much wider than the real interval. In order to reduce this IA drawback, the number of occurrences of the same parameter in the IA-based operations must be minimized. For instance, in the Equation (5), which allows one to calculate the PV generator current, θ includes the computation of an equivalent resistance resulting from the parallel between R_s and R_h; thus, $\frac{R_s \cdot R_h}{R_s + R_h}$. This expression involves two occurrences of each one of the resistances. By using the ranges $R_s = [0.01, 1]\Omega$ and $R_h = [500.5, 700.5]\Omega$, the IA gives the result $[0.00713471, 1.39957]\Omega$. By using the real arithmetic, obviously the equivalent expression $\frac{1}{\frac{1}{R_s} + \frac{1}{R_h}}$ gives the same results, but this is not true if IA is used. Indeed, a reduced number of variables' occurrences results in $[0.0099998, 0.998574]\Omega$. Therefore, the lower the number of occurrences of the interval valued parameters in (5), the more accurate the IA-based evaluation of the result. As widely shown in [33], this is not the only cause of overestimation of the interval of variation of the result of an operation by using IA, because the non linearity of the function operating over interval valued parameters and variables contributes to widening the resulting range.

In the PV-oriented problem treated in this paper, the PV current given by the SDM (5) is the explicit function $I = f([P], V)$ where $[P]$ is the interval valued vector of the parameters to identify and V is the real value of the voltage at which the current is evaluated. In case the set of interval values parameters is limited to two only, with the others being real values, the domain and co-domain of $[I] = f([P], V)$ is qualitatively depicted as in Figure 3. In the bi-dimensional plane representing the domain, $[P]_0$ is a initial square region in gray resulting from the two interval parameters $[P_1]_0$ and $[P_2]_0$. This corresponds to the co-domain $[I] = f([P]_0, V)$. A contraction of the domain, which is represented by a smaller rectangle resulting from the sets $[P_1]_1$ and $[P_2]_1$, corresponds to the co-domain $[I] = f([P]_1, V)$. If performed through the classical real valued analysis, e.g., by a Monte Carlo approach, the computations of the gray and the red envelopes in the I-V plane should require the selection of a high number of samples in the gray and the red rectangles in the parameters domain, and thus, a high number of Monte Carlo trials. The higher the number of $[P_1, P_2]$ couples, the more accurate the evaluation of the corresponding envelope in the I-V plane, which is obtained by merging all the curves obtained, and the voltage value by voltage value, by taking the maximum and the minimum I values. Such a computation should be able to reveal whether the experimental I-V curve is included or not in the envelope, and thus whether the corresponding sets $[P]_0$ or $[P]_1$ are feasible. A reliable evaluation of the envelopes, if performed by using the classical real numbers, thus, through a Monte Carlo method, should be more time consuming the more significant the non-linearity and non-monotonicity of the function I are with respect to the parameters. Instead, IA is a tool that allows one to evaluate the envelopes corresponding to each set $[P]_0$ and $[P]_1$ by a single computation. Thanks to the IA properties, the result will be guaranteed to bound the true I range.

In the next section, on the basis of such conclusions, the proposed IA parametric identification method is shown: it starts from a large rectangle in the parameters domain exemplified by the gray rectangle in the qualitative example of Figure 3, and contracts it in order to bound as much as possible the experimental points, which are marked in blue in Figure 3, in the I-V domain.

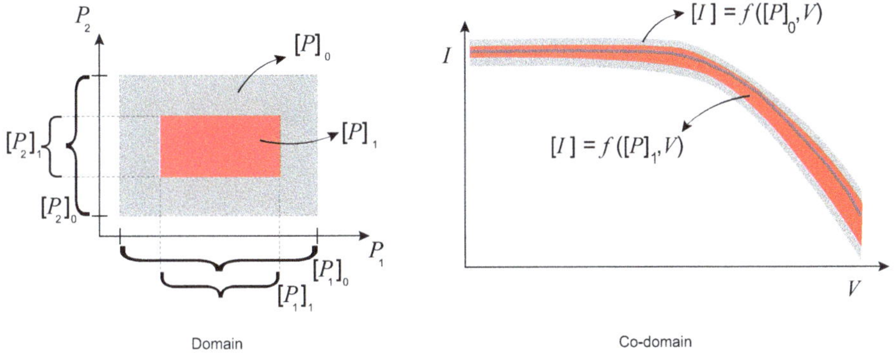

Figure 3. Domain and co-domain of interval function.

4. Parametric Identification by the IA-Based Divide-and-Conquer (*D&C*) Algorithm

In this paper, the identification of all the five parameters in (5) is considered. As a consequence, the rectangles shown in Figure 3 have to be considered hyper-rectangles in a 5-dimensional space. Iteration by iteration, the initial intervals $[P]$ are contracted in order to contract the $[I]$ around the experimental data. The iterations end when a termination condition fixed by the user is fulfilled. The divide-and-conquer (*D&C*) algorithm is an algorithm design paradigm for discrete and combinatorial optimization problems. The algorithm starts evaluating the largest candidate set of parameters assigned by the user, which is named $[P]_0$ in Figure 4. If the $[I]$ bounds include the experimental I-V samples (I_{exp}, V_{exp}), then all the five parameters intervals of $[P]_0$ are halved, so that 2^5 sub-intervals are generated.

For each of the 2^5 sub-intervals the I-V boundary is calculated by using IA. If the experimental points (I_{exp}, V_{exp}) are *not* included in the boundaries, the corresponding subset is marked as *infeasible* and it is not partitioned into smaller subsets anymore. On the contrary, the subset is partitioned by halving again the intervals it is made of and these ones are analyzed at the next iteration level. Thus, the algorithm continues with the next dividing level until a termination condition fixed by the user is fulfilled.

In summary, the proposed *D&C* algorithm consists of the following main elements appearing in Figure 4.

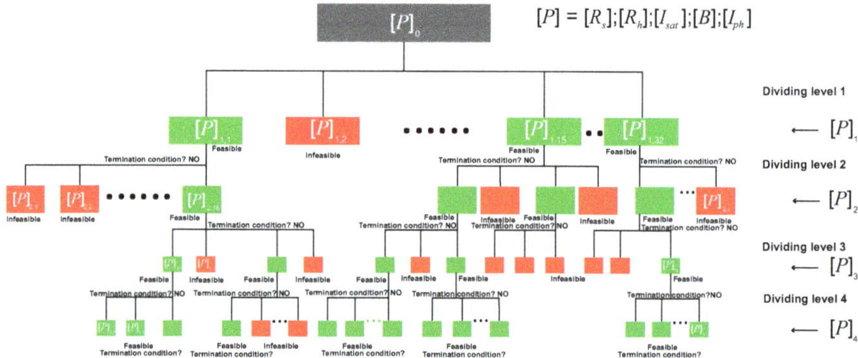

Figure 4. IA-based *D&C* approach.

Dividing Level (i): it is identified by the sub-index i. At each level, the parameters' intervals sets that fulfill the ***feasibility condition*** are halved, so that 2^5 new sub-intervals are generated. Thus, in case all the intervals are feasible, at the dividing level i a number 2^{5xi} is generated and its feasibility has to be tested. Each interval in the subset takes a new sub-index j, so that a particular set of parameters is called $[P]_{i,j}$. For instance, at the branching level $i = 1$, each element in the interval $[P]_0 = [[R_s]; [R_h]; [I_{sat}]; [B]; [I_{ph}]]$ is halved, and all the $2^{5x1} = 32$ combinations of these sub-intervals, which are called: $[P]_{1,1}...[P]_{1,32}$, have to be tested through the feasibility condition.

Feasibility condition: at the *i-th* branching level, the subset of intervals $[P]_{i,j}$ is substituted in (5) for each voltage value V_{exp} of the experimental data set. A current interval $[I]$ results at each voltage value and it is verified that the *all* the N_p experimental points fall within the calculated intervals:

$$I_{exp,n_p} \in [I([P], V_{n_p})], for\ all\ n_p = 1,\dots, N_p \tag{6}$$

Parameters' sub-intervals $[P]_{i,j}$ that do not fulfill the feasibility condition are not divided anymore and are not transferred to the next algorithm iteration.

It is worth noting that the infeasibility of these sub-intervals is guaranteed by the use of IA. Indeed, IA properties recalled in Section 3 ensure that the co-domain $[I] = f([P], V)$ evaluated over a set of parameters $[P]$ is an overestimation of the true range spanned by the current I for that domain $[P]$. As a consequence of the overestimation, if the range $[I]$ does not fulfill the feasibility condition, namely, does not include all the experimental I-V samples, then it is guaranteed to be infeasible. The same guarantee would be achieved by classical methods, e.g., Monte Carlo, only at a very high cost, even tending towards an infinite computational cost, thanks to the trials in the Monte Carlo approach. This represents a relevant advantage of the proposed IA based approach.

Termination condition: Feasible intervals $[P]_{i,j}$ falling below a minimum width, which is $wid[P]_{i,j}$, fixed by the user, are not divided further. Thus:

$$wid[P]_{i,j} \leq \Delta D \cdot mid[P]_{i,j} \tag{7}$$

where $mid[P]_{i,j}$ represents the midpoint of $[P]_{i,j}$. When the termination condition imposes that no more feasible intervals have to be partitioned further, then the union of the feasible intervals achieved by the algorithm represents the final result. The proposed IA-based *D&C* algorithm is presented in Algorithm 1.

Algorithm 1: IA-based *D&C* algorithm.

Data: $[P]_0 = \{[R_s]_0, [R_h]_0, [I_{sat}]_0, [B]_0, [I_{ph}]_0\}$: Initial Interval set of parameters

Result: $[P]_i = \{[R_s]_i, [R_h]_i, [I_{sat}]_i, [B]_i, [I_{ph}]_i\}$: Union of feasible intervals j of each parameter
      in the set, at a dividing level i that fulfill the termination condition

Initialization: $i = 0, j = 1, \Delta D = 0.015$;

The five parameters intervals $[P]_i$ are halved, thus $i = 1$ and $j = 1, ..., 32$, obtaining the set:
  $[P]_{i,j} = \{[P]_{1,1}...[P]_{1,32}\}$;

Calculate the width and midpoint of the intervals: $wid[P]_{i,j}; mid[P]_{i,j}$ for all i, j ;

while $(wid[P]_{i,j} \geq \Delta D \cdot mid[P]_{i,j}$ for all $i, j)$ **do**

  if $I_{exp,n_p} \in [I([P]_{i,j}, V_{n_p})]$, for all i, j and all $n_p = 1, ..., N_p$ **then**

    Each $[P]_{i,j}$ is halved, i++;

    The maximum value of j, J_{max}, depends on the number of feasible intervals;

  else

    Infeasible intervals $[P]_{i,j}$ are discarded;

  end

end

Union $\cup$ of feasible intervals $[P]_{i,j}$, at the maximum dividing level i, for all $j = 1..., J_{max}$:

$[R_s]_i = [R_s]_{i,1} \cup [R_s]_{i,2} \cup ...[R_s]_{i,J_{max}}$; $[R_h]_i = [R_h]_{i,1} \cup [R_h]_{i,2} \cup ...[R_h]_{i,J_{max}}$;

$[I_{sat}]_i = [I_{sat}]_{i,1} \cup [I_{sat}]_{i,2} \cup ...[I_{sat}]_{i,J_{max}}$; $[B]_i = [B]_{i,1} \cup [B]_{i,2} \cup ...[B]_{i,J_{max}}$;

$[I_{ph}]_i = [I_{ph}]_{i,1} \cup [I_{ph}]_{i,2} \cup ...[I_{ph}]_{i,J_{max}}$;

---

5. Identification of the Parameters R_s, R_h and I_{sat} through the IA-based *D&C* Method

The identification of the SDM parameters almost consists of identifying R_s, R_h and I_{sat}. Indeed, the I_{ph} parameter is assumed as equal to the short-circuit current [2], whose value is experimentally measured. On the other hand, once having measured the cells' temperature and by assuming that the number of series connected cells in the module is known, the value of the parameter B is fixed if, as it is quite common in literature, (see [8,34,35]), n assumes a value between 1 and 2. Typical values are below 1.5, but the search range has been extended up to 2 in order to account for more extreme cases documented in literature [36]. It has to be kept in mind that the range is subjected to the contraction due to the IA based approach proposed in this paper. Thus, a smaller upper limit would not affect the final result of the identification process, but its rate of convergence. With those assumptions, the identification process limited to the three parameters R_s, R_h and I_{sat} is of practical interest and allows one to demonstrate the performance of the proposed *D&C* algorithm on a reduced scale case. The algorithm in this case is tested by using I-V samples that are obtained by using the parameters in Table 2 in the SDM (5). Samples are calculated at a fixed voltage step $0.1V$, so that the $N_p = 222$ I-V samples shown in Figure 5 are considered.

The *D&C* algorithm has run on the following search space: $[R_s] = [0.1, 1]\Omega; [R_h] = [1, 1000]\Omega; [I_{sat}] = [1e^{-8}, 1e^{-12}]A$. The nominal values for B and I_{ph} given in Table 2 have been also used. $\Delta D = 1.5\%$ has been used for settling the termination condition.

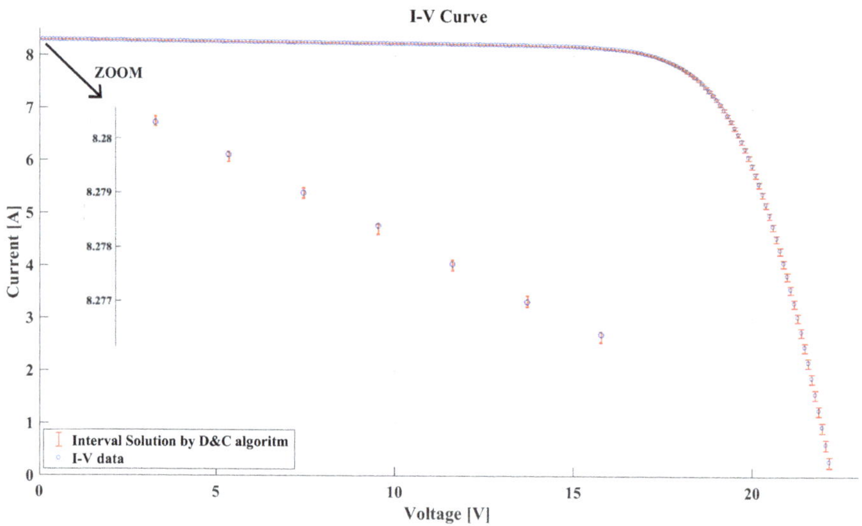

Figure 5. Performance of *D&C* by using noiseless I-V simulated data.

The algorithm has created the number of dividing levels shown in Table 3. The third column of Table 3 reveals the effectiveness of the proposed IA approach. Indeed, as pointed out before, the main advantage of applying IA to the feasibility condition is the immediate and guaranteed classification of the infeasible sets of the search space. It is evident that, for this example, just at the first dividing level, 50% of the search space is immediately classified as infeasible. The same result would require a number of Monte Carlo trials instead of four IA based computations. The fourth column of Table 3 gives a measure of the volume of each subset at the corresponding dividing level. The solution is reached at the tenth dividing level, at which two sets of interval solutions have been identified. The union of those two intervals is shown in Table 4. It reveals that the interval set solution contains the values of parameters R_s, R_h and I_{sat} used to generate the I-V samples. This is the expected result, so that the convergence property of the *D&C* algorithm is confirmed. A personal computer (PC) equipped with a Corei7-3632QM processor @ 2.20 GHz, four cores and 8 GB of RAM memory is used. The executable file, produced by starting from the C++ source, was run on a PC. With this software and hardware, the algorithm reaches the solution in 1.02 s after 360 iterations. Figure 5 puts into evidence that all the I-V samples fall inside of interval current $[I]$ determined by the IA based method.

This first test has been done by identifying the parameters by using I-V samples obtained through the same model, the SDM, adopted for the identification thereof. In this way, the process has not been affected by inaccuracies of the SDM in fitting experimental data and by inaccuracies and noise over I-V measurements. These effects will be more evident in the next sections wherein experimental I-V data are used.

Table 3. Number of feasible intervals, percentage of infeasible intervals and volume of the subsets at each dividing level for the example using noiseless data.

Dividing Level	Number of Feasible Intervals	% of Infeasible Intervals	Subsets Volume
1	4	50.00%	1.2361×10^{-6}
2	9	71.88%	1.5452×10^{-7}
3	3	95.83%	1.9315×10^{-8}
4	4	83.00%	2.4143×10^{-9}
5	5	84.38%	3.0179×10^{-10}
6	4	90.00%	3.7724×10^{-11}
7	6	81.25%	4.7155×10^{-12}
8	4	91.60%	5.8944×10^{-13}
9	5	84.38%	7.3680×10^{-14}
10	2	95.00%	9.2100×10^{-15}

Table 4. Interval solution by *D&C* algorithm in the example using noiseless data.

Parameters	Initial Intervals	Union of Intervals in the Space of Solutions
$[R_s]$	$[0.1,1]\ \Omega$	$[0.179189, 0.180156]\ \Omega$
$[R_h]$	$[1,1000]\ \Omega$	$[150.265, 151.24]\ \Omega$
$[I_{sat}]$	$[1 \times 10^{-12}, 1 \times 10^{-8}]\ \text{A}$	$[5.28291 \times 10^{-10}, 5.4782 \times 10^{-10}]\ \text{A}$

6. *D&C* IA-Based Approach Applied to Experimental Data

Experimental I-V data are commonly affected by noise due to, e.g., sensor quality and the data acquisition system's resolution. The low voltage region usually is the most critical because it requires a high resolution of the current sensor. Similarly, although not so seriously as in the previous case, at low current the voltage sensor has to show a significant resolution. In the presence of noise, such critical aspects become more and more significant. Therefore, the proposed *D&C* method has been made more robust in order to cope with noisy experimental data. Firstly, the decision on whether the experimental value is within or outside the IA determined $[I]$ interval is taken by account for a suitably small noise band around the experimental value. Additionally, the feasibility condition is relaxed by considering that an interval set of parameters is feasible if a number, but not all, of the experimental data fulfill the condition (6). The effects of these two additional conditions are analyzed in detail in the following subsections.

6.1. Relaxation of the Inclusion Property: First Approach

In Figure 6, the blue circles, which are the experimental data, are surrounded by blue bars representing a noise band, named $[I_{exp}]$. The red bars bound the interval of the current $[I]$, which is computed by IA on a given interval parameters set. Thus, the inclusion property is reformulated by considering that the experimental value (blue dots) is included in the interval range (red interval) if the intersection between the red range and the blue range of that experimental point is not empty. The larger the noise or the uncertainty, e.g., related to the sensors used, affecting the experimental data, the wider the band $[I_{exp}]$ and the higher the probability that the intersection between $[I_{exp}]$ and $[I]$ is not empty, and thus that the corresponding set of parameters is feasible.

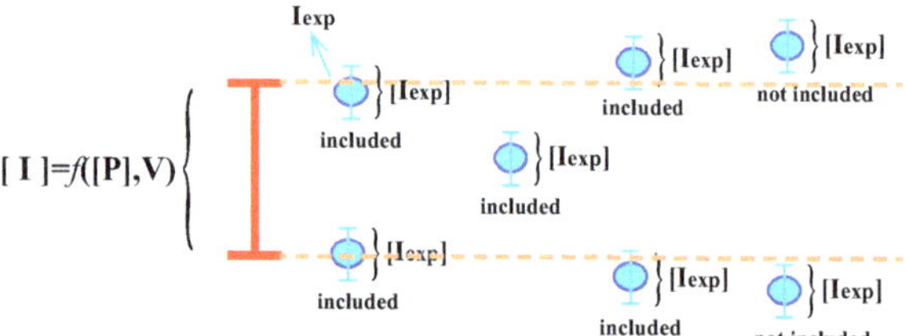

Figure 6. Inclusion property of experimental data.

6.2. Relaxation of the Inclusion Property: Second Approach

In (8), the feasible condition that relaxes the (6) by marking as feasible a parameter set for which $N_p^- < N_p$ experimental points fall within the I-V boundaries is formalized. This approach allows one to restrict the application of the feasibility condition to some regions of the I-V curve wherein the major information content is concentrated. For instance, the experimental I-V samples that are more affected by measurement noise can be excluded, or the feasibility condition can be limited to a region including the MPP, to the short circuit and to the open circuit conditions.

$$[I_{exp,n_p}] \in [I([P], V)], for\ n_p = 1, \ldots, N_p^- \tag{8}$$

6.3. D&C Parametric Identification by Using Noisy Experimental Data

Figure 7 shows an experimental set of I-V data referring to a 140W PV Yingli solar module that have been acquired at an irradiance equal to 849 W/m^2 and at a temperature equal to 336.15 K: they have been measured by using a low cost system described in detail in [37], which has the drawback of providing a high number of data in the low current range and a low density of measurements at low voltages. The I-V curve is obtained by the capacitor charging method, which takes an acquisition time of 0.05 s. The experimental setup is described in detail in [37]: the PV module output is made available in the laboratory through a 10 m long cable. Thus, the acquired I-V curve also takes into account the parasitic resistance of the cables, which is 60 mΩ. Figure 7 shows that the module under test has suffered degradation, due to 3.5 years of long operation.

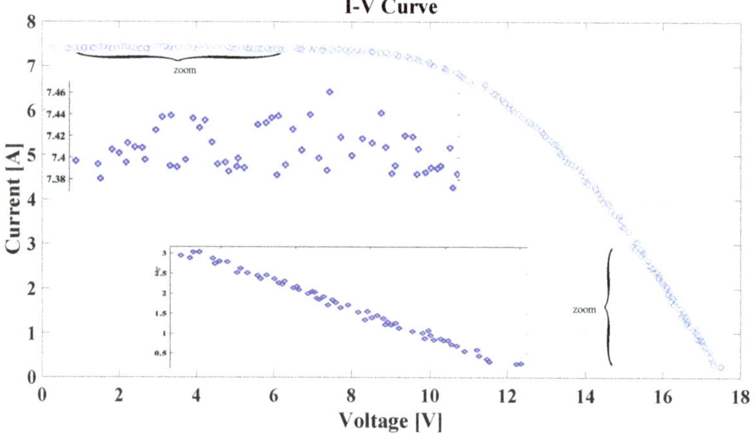

Figure 7. Experimental I-V curve.

6.3.1. Identifying $[R_s]$, $[R_h]$, $[I_{sat}]$, $[B]$ and $[I_{ph}]$

The parameter N_p^- in (8) has been settled to 216, so that 90% of experimental points have been taken into account for the feasibility condition. Thus, 10% of points are excluded, regardless of their position in the I-V curve, but the short-circuit current I_{sc} and the open circuit voltage V_{oc} have been always included in the 90%, so that the feasibility condition for this example has been formalized as in (9):

$$[I_{exp,n_p}] \in [I([P], V_{n_p})], for \ n_p = 1, \ldots, N_p^-$$
$$[I_{exp,1}] \in [I([P], V_1)]$$
$$[I_{exp,N_p}] \in [I([P], V_{N_p})]$$

(9)

The current samples at low voltage show noise close to 1% of their values; thus, the parameter appearing in the termination condition ΔD is fixed at 10%. Table 5 collects the number of feasible intervals at each division level. Additionally, in this example, from the third column of Table 5, it comes out that the IA approach guarantees the infeasibility of 56.25% of the initial search space at the first dividing level. Indeed, of the initial $2^5 = 32$ subsets, 18 are guaranteed to be infeasible by a direct computation of $[I]$ through IA. These results reveal that the IA-based $D\&C$ proposed algorithm finds the solution at the dividing level 5, in which there are 5817 interval sets that have been classified feasible. The algorithm spent 11.68 min to run 242,496 iterations. It is worth noting that these numbers are significantly higher than those ones achieved in the example presented in Section 5. This is due to the presence of noise, affecting the experimental samples and not the simulated samples of the previous case, and it is also a consequence of the chosen value of the termination threshold, which is now fixed at $\Delta D = 10\%$, and thus greater than $\Delta D = 1.5\%$ used in the previous example. The union of the feasible intervals sets achieved at the last dividing level is given in the third column of Table 6. The contraction of the intervals with respect to the initial search space has been also shown.

Table 5. Number of feasible intervals, percentage of infeasible intervals and volume of the subsets at each dividing level for the example using experimental I-V data.

Dividing Level	Number of Feasible Intervals	% of Infeasible Intervals	Subsets Volume
1	14	56.25%	8.3219×10^{-4}
2	148	66.96%	2.6006×10^{-5}
3	1089	77.00%	8.1269×10^{-7}
4	6326	81.85%	2.5396×10^{-8}
5	5817	97.13%	7.9364×10^{-10}

Table 6. Interval solution by the $D\&C$ algorithm at the final dividing level in the example using experimental I-V data.

Parameters	Initial Intervals	Union of Feasible Intervals
$[R_s]$	$[0.1, 1]\Omega$	$[0.26875, 0.60625]\Omega$
$[R_h]$	$[300, 800]\Omega$	$[300, 800]\Omega$
$[I_{sat}]$	$[1 \times 10^{-8}, 1 \times 10^{-4}]$A	$[3.12569 \times 10^{-5}, 7.81272 \times 10^{-5}]$A
$[B]$	$[0.7, 1.5]$	$[1.375, 1.5]$
$[I_{ph}]$	$[7.0271, 7.7669]$A	$[7.397, 7.42012]$A

In Figure 8, red bars give the current intervals calculated by substituting the interval solution of Table 6 in (5) and using IA. At least 90% of experimental data, those in blue marks, fall inside the interval current $[I]$. Nevertheless, although a significant contraction of the search space has been achieved by the $D\&C$ method (see Table 6), the interval solution (red bars) still gives a large range around the experimental points. As it will be shown in the next subsection, an improved identification accuracy is achieved by analyzing the feasible sub-intervals achieved by the *D&C* method at a given dividing level.

The identified intervals achieved and shown in Table 6 can be used to identify, in turn, some additional SDM parameters. For instance, the interval $B = [1.375, 1.5]$ can be used to identify the corresponding interval of the diode ideality factor n, again by using IA. Indeed, it results that:

$$n = \frac{B \cdot q}{N_s \cdot k \cdot T} = \frac{[1.375, 1.5] \cdot 1.60217662 \cdot 10^{-19}}{36 \cdot 1.38064852 \cdot 10^{-23} \cdot 336.15} = [1.31854, 1.43841] \tag{10}$$

Additionally, the uncertainty of the devices used in the temperature measurement system can account for: a LM35 installed on rear side of the PV module, a non-inverting amplifier and a 10 bit ADC. In this case, the uncertainty affecting the temperature measurement is equal to 0.8%. Thus, it results that:

$$n = \frac{[1.375, 1.5] \cdot 1.60217662 \cdot 10^{-19}}{36 \cdot 1.38064852 \cdot 10^{-23} \cdot [333.461, 338.839]} = [1.30808, 1.45001] \tag{11}$$

Both the intervals achieved for n are in a suitable range for this parameter. The same procedure might be applied by identifying further physical parameters underlying the set of five shown in Table 6.

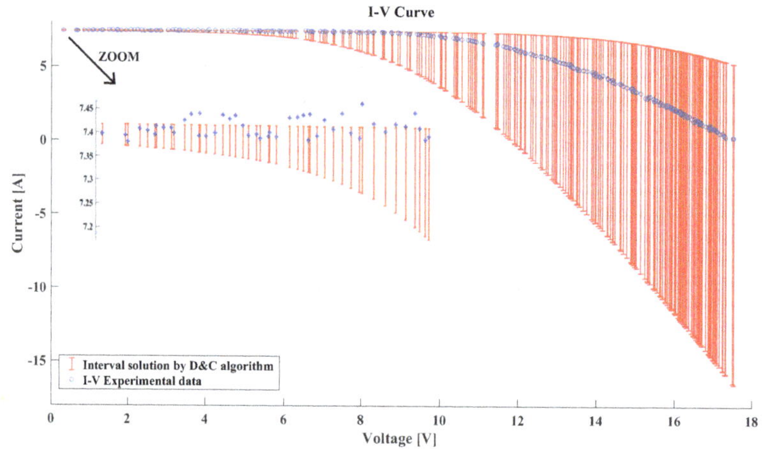

Figure 8. Experimental I–V curve vs. SDM using the interval solution set of Table 6.

6.3.2. Analysis of the Feasible Sub-Intervals

In the previous example, the final solution obtained through the IA based algorithm was determined as the union of all the feasible sets of intervals achieved at the final dividing level. Each sub-interval can be examined in more detail by calculating the current obtained, at each voltage value of the experimental samples, through SDM (5) using the midpoints $mid[P]$ of the interval of each parameter. This calculated current is called $I(mid[P], V)$. Then, the root-mean-square error (RMSE) of the identified current with respect to the experimental I_{exp} is calculated. In (12), the RSME formula is shown, which takes into account the number of samples N_p. The sub-interval giving the I-V curve having the minimum value of the RMSE is considered as the best. This analysis has been applied to the feasible sub-intervals in Table 5, so that 5817 set of intervals have been evaluated. The second column of Table 7 shows the $mid[P]_{5,best}$ found in the 5817 sets in the space solutions, which correspond to the interval set $[P]_{5,4350}$. The fit of the I-V curve generated using $mid[P]_{5,best}$ is presented in Figure 9. The minimum RMSE value achieved is 0.0659.

$$RMSE = \sqrt{\frac{\sum_{i=1}^{N_p}(I_{exp} - I(mid[P], V))^2}{N_p}} \tag{12}$$

Table 7. $mid[P]_{best}$ and the corresponding interval solution with the smallest RMSE value.

Parameters	$mid[P]_{5,best}$	Best Interval Solution
R_s	0.4797 Ω	[0.4656, 0.4938] Ω
R_h	792.1875 Ω	[784.3750, 800] Ω
I_{sat}	4.5318×10^{-5} A	[4.3756×10^{-5}, 4.6880×10^{-5}] A
B	1.4625	[1.4500, 1.4750]
I_{ph}	7.4086 A	[7.3970, 7.4201] A

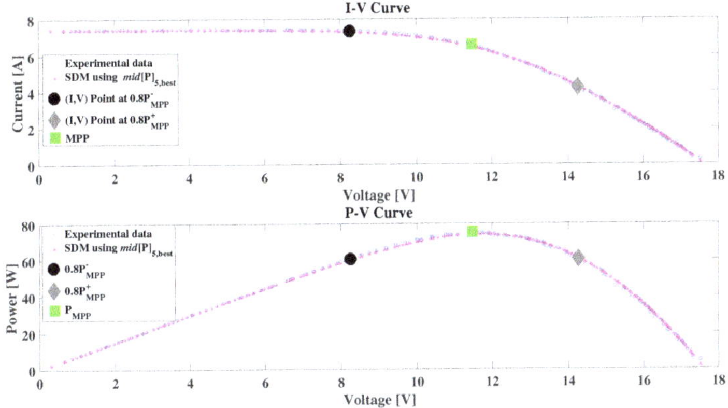

Figure 9. I-V curve: experimental data vs. SDM using $mid[P]_{5,best}$.

In Figure 10, the I-V curve generated by SDM using the best sub-interval, which is shown in the third column of Table 7, is depicted: the contraction of the initial interval set with respect to the Figure 8 is evident.

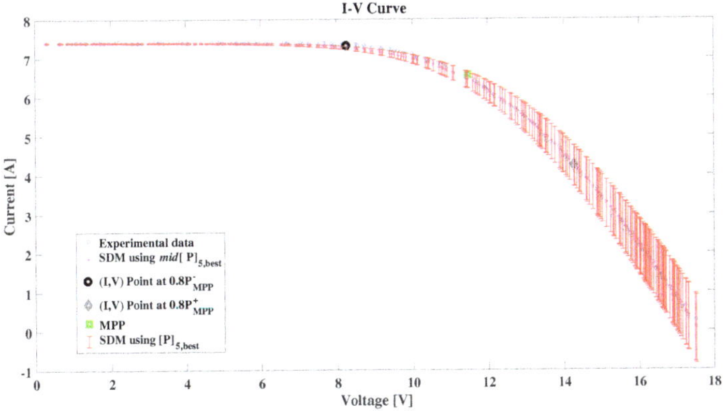

Figure 10. I-V curve: experimental vs. SDM using the interval solution.

6.3.3. Identifying $[R_s]$, $[R_h]$, $[I_{sat}]$, $[B]$ and $[I_{ph}]$ by the D&C Algorithm with Emphasis Around the MPP

The experimental points of the I-V curve across its MPP are of primary importance for the parametric identification of the SDM, and they might be excluded by the procedure described above. In this example, the experimental data around the MPP and limited to the range [80%, 100%] of the power at the MPP, which is P_{MPP}, are considered. The left and right extremes are shown in Figures 9–11 in black and gray, and are named $0.8P_{\mathrm{MPP}}^-$ and $0.8P_{\mathrm{MPP}}^+$. Those power values correspond to the samples called $N_p^{0.8P_{\mathrm{MPP}}^-}$ and $N_p^{0.8P_{\mathrm{MPP}}^+}$, respectively. All the experimental points in this range have been included in the feasibility condition (13). The tolerance around experimental data I_{exp} and the termination condition ΔD have been fixed at 1% and 10% respectively.

Relaxed feasibility condition with constraints of I_{sc} and V_{oc} and less data.

$$[I_{exp,n_p}] \in [I([P], V_{n_p})], for \; n_p = N_p^{0.8P_{\mathrm{MPP}}^-}, \ldots, N_p^{0.8P_{\mathrm{MPP}}^+}$$

$$[I_{exp,1}] \in [I([P], V_1)]$$

$$[I_{exp,N_p}] \in [I([P], V_{N_p})] \tag{13}$$

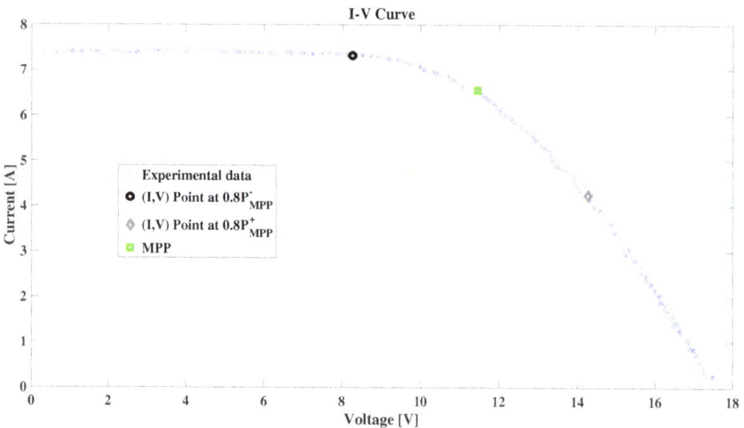

Figure 11. Maximum power point (MPP) and points ensuring the 80% of P_{MPP} in the experimental I-V curve.

The results of D&C algorithm are shown in Table 8. Again, in this example, at the first dividing level, IA classifies as infeasible 18 subsets over 32, just by a direct IA based computation of the current interval $[I]$. The solution is reached at the dividing level 5, in which the space of solutions contains 2890 sub-intervals. In this case the algorithms needs 213,952 iterations; thus, the computation time and memory are reduced with respect to the previous examples. The number of feasible sub-intervals in the final dividing level is reduced by 51.4% and the number of iterations is reduced by 11.7%. The union of the final sub-intervals is given in Table 9, and it reveals the contraction with respect to the initial search space. The set of intervals is similar to the one obtained by using all the experimental values; R_s is the only one showing an improved contraction.

Table 8. Number of feasible intervals, percentage of infeasible intervals and volume of the subsets at each dividing level for the example using a reduced set of experimental data.

Dividing Level	Number of Feasible Intervals	% of Infeasible Intervals	Subsets Volume
1	14	56.25%	8.3219×10^{-4}
2	144	67.86%	2.6006×10^{-5}
3	1032	77.60%	8.1269×10^{-7}
4	5495	83.36%	2.5396×10^{-8}
5	2890	98.36%	7.9364×10^{-10}

Table 9. Interval solution by the *D&C* algorithm in the example using a reduced set of experimental data.

Parameters	Initial Intervals	Union of Feasible Intervals
$[R_s]$	$[0.1,1]\Omega$	$[0.325,0.55]\Omega$
$[R_h]$	$[300,800]\Omega$	$[300,800]\Omega$
$[I_{sat}]$	$[1 \times 10^{-8}, 1 \times 10^{-4}]$A	$[3.12569 \times 10^{-5}, 7.81272 \times 10^{-5}]$A
$[B]$	$[0.7,1.5]$	$[1.375,1.5]$
$[I_{ph}]$	$[7.0271,7.7669]$A	$[7.397,7.42012]$A

In Figure 12, the red bars correspond to SDM evaluated by IA for the solution presented in Table 9. As in the previous case, large ranges result from the union of the feasible sub-intervals at the final dividing level where the algorithm terminated. The analysis of the RMSEs for all the feasible sub-intervals at the division level 5 gives a narrower range. The best sub-interval is the same as that achieved in the previous example, and is shown in Table 7 and Figure 10. The best sub-interval set is $[P]_{5,2272}$, and the minimum RMSE value is 0.0776.

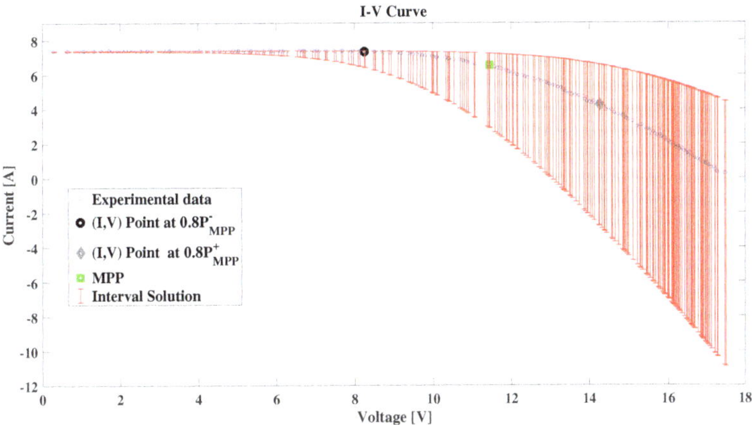

Figure 12. I-V curve: experimental vs. SDM using *D&C* algorithm with a reduced set of experimental data

7. Discussion of the Results

Some aspects concerning the results presented in the previous examples deserve further comments. The first one concerns the way in which the initial interval set of parameters, and thus the search space, is chosen. The proposed IA-based *D&C* algorithm was run on an initial interval set $[P]_0$ that was generally very large, just in order to test the convergence and contraction capabilities of the approach. In the first example, which referred to the identification of the values of three parameters only and used I-V data generated by the same model used for the identification thereof, a large initial interval set was used. The initial intervals for the two resistances were set to include typical values; thus, they were in the order of magnitude of hundreds of mΩ and hundreds of Ω for series and parallel resistances respectively. Such initial intervals might also include values corresponding to a degraded PV module. The initial interval of I_{ph} has been chosen across the short-circuit current value I_{sc}.

By using the I-V experimental measurements around the MPP, the I_{ph} is settled at values that are close to the MPP current I_{mpp}; thus, an initial interval across I_{mpp} is used. The initial ranges of B and I_{sat} have been determined by keeping account of some physical relationships. The parameter B depends on temperature T and n: it has been assumed that T has been measured with a known accuracy and that the ideality factor, as can be deduced from the literature referring to silicon cells, assumes values ranging from $n = 1$ up to $n = 2$. The larger the uncertainty affecting the measure of the temperature, the wider the initial interval $[B]_0$. As for I_{sat}, it has been assumed that, for new PV modules, it assumes values of the order of nA while μA is used for aged modules. The $[I_{sat}]_0$ width affects the convergence features of the approach significantly. The proposed examples show the convergence capability of the D&C algorithm even using a $[I_{sat}]_0$ that is four orders of magnitude and has n ranging up to 2, instead of stopping at 1.3, as can be deduced by reading some papers; e.g., [35]. However, a better trade-off between accuracy and computation time should be reached by having a more accurate estimation of the initial range of the parameters.

The second aspect deserving further comments concerns the selection of the value ΔD involved in the termination condition (7), because it affects the accuracy of the result and the computation time required by the algorithm. In the case a low noise level affecting the I-V samples, a tiny termination condition does not affect the computation time significantly, as in the first example. Indeed, any relaxation of the inclusion property is required and a small number of feasible intervals at each dividing level is obtained. In the case of I-V experimental data exhibiting a significant noise level, a trade-off between accuracy and computation time needs to be achieved. Some relaxation of the inclusion property and a higher value of ΔD help to achieve the convergence. It is worth noting that the number of feasible subsets obtained at the end of each algorithm run depends on both the ability of the SDM to fit the experimental curve and on the chosen ΔD value. The additional step using the RMSE calculation discussed in some examples presented in Sections 5 and 6 helps to improve the accuracy of the IA solution.

The third remark concerns the size of interval current $[I]$, as it is shown in Figures 8, 10 and 12. In the SDM solution shown in Figure 8 and Table 6, the relative width of the interval parameters' solution ($wid_m[P]$), is calculated by $\frac{wid[x,\bar{x}]}{mid[x,\bar{x}]}$. The results are $wid_m[R_s] = 0.7714$, $wid_m[R_h] = 0.9091$, $wid_m[I_{sat}] = 0.8570$, $wid_m[B] = 0.0870$ and $wid_m[I_{ph}] = 0.0031$. Figure 12 shows the I-V curve boundaries corresponding to the same interval solution, by neglecting the range of $[R_s]$. In this case, the relative width is 0.5143; thus, the important effect of the $[R_s]$ interval on $[I]$ becomes evident. By using the RMSE calculation in Table 7, the relative interval sizes are reduced to the following values: $wid_m[R_s] = 0.0588$, $wid_m[R_h] = 0.0197$, $wid_m[I_{sat}] = 0.0689$, $wid_m[B] = 0.0171$ and $wid_m[I_{ph}] = 0.0031$. The significant effect of this contraction on the range $[I]$ is evident by looking at Figure 10. The contraction is close to one order of magnitude for all the parameters, but not for $[I_{ph}]$. Figure 2 shows that $[I_{sat}]$ and $[B]$ have significant effects on the $[I]$ width. Figure 13 shows that the true range of $[I]$ is overestimated because of the use of IA, especially at high voltage. The overestimation is evident by comparing the IA results with those ones obtained by means of a Monte Carlo run over 2000 random trials. The corresponding I-V curves are shown in black color, which have been generated by randomly choosing sets of parameters in the ranges shown in Table 7. It is worth noting that the Monte Carlo method giving a narrower range with respect to the IA method does not mean that the former result is more accurate than the latter one. Indeed, only if both of them are taken into account, exact information about the true range spanned by I at the different voltages is obtained. Indeed, the Monte Carlo range would approach the true one by running an infinite number of trials; otherwise it gives an underestimation of the true range of I. The IA overestimation is reduced by reducing the width of the interval parameters [33]. The true range is placed in the middle, bounded by the Monte Carlo range, which is an underestimation of the true range, and the IA range, which is an overestimation of the true range.

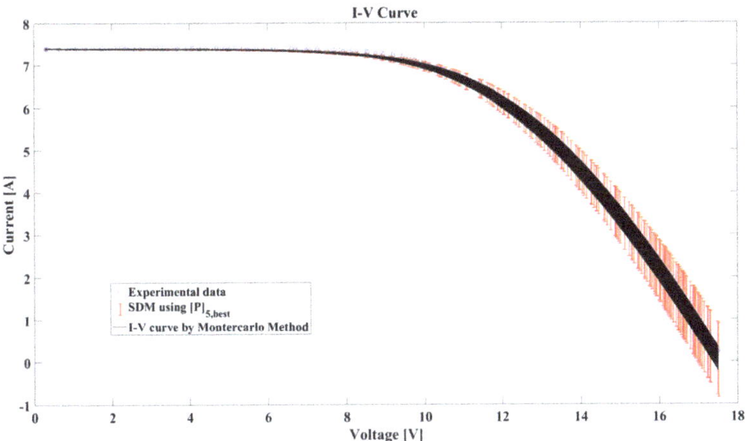

Figure 13. Monte Carlo analysis of interval solution.

An additional advantage of the proposed IA-based *D&C* algorithm can be put into evidence by referring to the results shown in Table 10. The *Matlab Fit APP* tool has been used to identify the five parameters of SDM. It minimizes the root mean square error between the experimental I-V data and I-V curve obtained through the SDM with the identified values of the parameters. The trust-region method has been selected, the function tolerance value has been settled at 1e-5 and the maximum number of iterations has been fixed at 400. In the second column the results achieved by this tool are given. The initial interval of the parameters has been set equal to the initial interval used in the proposed IA-based *D&C* algorithm; thus, the one given in the second column of Table 6. The third column of Table 10 shows the result when the initial search space used in *Matlab Fit APP* is the union of the feasible intervals obtained by the the proposed IA-based *D&C* algorithm; thus, the one in the third column of Table 6. It is evident that the proposed IA based approach has contracted the initial search space towards the solution in an effective way, so that the *Matlab Fit APP* converges to the identified set by a number of iterations and function evaluations that is 80% lower than the one required if the search is started from the wider search space used by the IA *D&C* method. Moreover, the step size is reduced by four orders of magnitude, so that a higher accuracy in the parameter identification is achieved. This result reveals that the feasible intervals obtained by the proposed IA-based *D&C* algorithm are reliable guess solutions for gradient based minimization methods. The cascade of the methods thus allows one to improve the convergence and the accuracy of the result. The RMSE value obtained by *Matlab Fit APP* is equal to 0.0587, which is close to the one obtained by the proposed analysis procedure of the feasible sub-intervals ($mid[P]_{5,best}$ shown in Table 7), which is 0.0659. The *D&C* IA-based method uses a simple partitioning of the intervals and feasibility test, and thus, any gradient or minimization method, also guaranteeing the infeasibility of the discarded intervals.

Table 10. Performance comparison of *Matlab Fit APP* tool using the intervals of Table 6.

Feature	Using the Initial Intervals	Using the Union of Feasible Intervals	Improvement
Number of iterations	67	13	80%
Number of function evaluations	408	84	79%
Step Size	0.0707	5.2217×10^{-6}	Four orders of magnitude

As a further comment concerning the implementation of the IA-based *D&C* algorithm, it has to be evidenced that it might profit significantly from a parallel implementation of the IA operations. Indeed, in any IA operation, the computations of the lower bound and of the upper bound of the result can be done in parallel, because these two computations are independent. Moreover, the computation tree derived from the proposed *D&C* method is also prone to a parallel computation. Consequently, the proposed algorithm, which has been already developed in C++ by means of a suitable library including all the IA operations, can be implemented in embedded devices including multi-core processors or field programmable gate arrays (FPGAs).

8. Conclusions

In this paper an interval-arithmetic-based approach has been applied for identifying the five parameters of the single-diode model of crystalline silicon photovoltaic modules. The divide-and-conquer computational paradigm has been used to contract an initial interval set of parameters, that is, the search space, towards an interval parameters set of a suitably small width. The proposed method generates feasible and infeasible intervals by successive divisions of the initial search space. Each interval is evaluated by a feasible condition through interval arithmetic: this key operation allows one to discard infeasible portions of the search space with a single operation, without involving any iterative procedure or any minimization algorithm. Moreover, interval arithmetic guarantees the infeasibility of the discarded sets, meaning that no combinations of parameters in those sets give a current vs. voltage curve that is more close to the experimental samples than the curves obtained by the feasible sets. After discarding the infeasible intervals, the proposed method reduces the widths of the feasible ones until they fall below a threshold fixed by the user through the termination condition. The performance of the proposed algorithm has been tested on three examples, including simulated data and experimental data, the latter affected by measurement noise. The analysis of the case using experimental measurements has evidenced the need for a further computation step that profits from the interval contraction capabilities of interval arithmetic, allowing one to refine the final interval solution. In addition to the main feature of parametric identification, the proposed algorithm gives some information that should be useful in the detection of aging, malfunctioning and faults of the photovoltaic generator. Indeed, the final result of the application of the method gives an indication about the sensitivity of the model with respect to the five parameters appearing in it. Moreover, the ranges provided by the method and including the experimental current vs. voltage samples give a mask for linking the variation of the module performance to the variation of its parameters. Thanks to the interval arithmetic inclusion properties, current values acquired in the same operating conditions and falling outside the interval ranges would reveal variations of the parameters that are outside the corresponding ranges. For instance, by applying this on-site evaluation to the series resistance, the aging of the module exceeding a fixed threshold can be detected. The offline computation of the interval boundaries in the current vs. voltage plane and their uploading on a low cost processor would allow a straightforward and on-site verification of the violation of these boundaries with negligible computation effort.

Funding: This work has been supported by Universidad del Valle, Cali, (Colombia), under the project CI 1036. Moreover, the author gratefully acknowledge the financial support provided by the Colombia Scientific Program within the framework of the call Ecosistema Científico (contract number FP44842- 218-2018).

Acknowledgments: This work has been supported by Universidad del Valle, Cali, (Colombia), and the Colombia Scientific Program within the framework of the call Ecosistema Científico.

Conflicts of Interest: The author declares no conflict of interest.

References

1. Suthar, M.; Singh, G.; Saini, R. Comparison of mathematical models of photo-voltaic (PV) module and effect of various parameters on its performance. In Proceedings of the 2013 International Conference on Energy Efficient Technologies for Sustainability, Nagercoil, India, 10–12 April 2013.

2. Petrone, G.; Ramos-Paja, C.; Spagnuolo, G. *Photovoltaic Sources Modeling*; Wiley-IEEE Press: Hoboken, NJ, USA, 2017.

3. Hare, J.; Shi, X.; Gupta, S.; Bazzi, A. Fault diagnostics in smart micro-grids: A survey. *Renew. Sustain. Energy Rev.* **2016**, *60*, 1114–1124. [CrossRef]

4. Mesquita, D.; Silva, J.; Moreira, H.; Kitayama, M.; Villalba, M. A review and analysis of technologies applied in PV modules. In Proceedings of the IEEE PES Innovative Smart Grid Technologies Conference—Latin America (ISGT Latin America), Gramado, Brazil, 15–18 September 2019.

5. IRENA. *Future of Solar Photovoltaic-Deployment, Investment, Technology, Grid Integration and Socio-Economic Aspects (A Global Energy Transformation: Paper)*; International Renewable Energy Agency: Abu Dhabi, UAE, 2019.

6. NREL. *Best Research-Cell Efficiency Chart*; National Renewable Energy Laboratory: Golden, CO, USA, 2019.

7. Tossa, A.K.; Soro, Y.; Azoumah, Y.; Yamegueu, D. A new approach to estimate the performance and energy productivity of photovoltaic modules in real operating conditions. *Sol. Energy* **2014**, *110*, 543–560. [CrossRef]

8. Bai, J.; Liu, S.; Hao, Y.; Zhang, Z.; Jiang, M.; Zhang, Y. Development of a new compound method to extract the five parameters of PV modules. *Energy Convers. Manag.* **2014**, *79*, 294–303. [CrossRef]

9. Mares, O.; Paulescu, M.; Badescu, V. A simple but accurate procedure for solving the five-parameter model. *Energy Convers. Manag.* **2015**, *105*, 139–148. [CrossRef]

10. Bishop, J.W. Computer Simulation of the Effects of Electrical Mismaches in Photovoltaic Cell Interconnection Circuits. *Sol. Cells* **1998**, *25*, 73–89. [CrossRef]

11. Chin, V.J.; Salam, Z.; Ishaque, K. Cell modelling and model parameters estimation techniques for photovoltaic simulator application: A review. *Appl. Energy* **2015**, *154*, 500–519. [CrossRef]

12. Batzelis, E. Non-Iterative Methods for the Extraction of the Single-Diode Model Parameters of Photovoltaic Modules: A Review and Comparative Assessment. *Energies* **2019**, *12*, 358. [CrossRef]

13. Cannizzaro, S.; Di Piazza, M.C.; Luna, M.; Vitale, G. PVID: An interactive Matlab application for parameter identification of complete and simplified single-diode PV models. In Proceedings of the 2014 IEEE 15th Workshop on Control and Modeling for Power Electronics (COMPEL), Santander, Spain, 22–25 June 2014; pp. 1–7. [CrossRef]

14. Villalva, M.G.; Gazoli, J.R.; Filho, E.R. Comprehensive Approach to Modeling and Simulation of Photovoltaic Arrays. *IEEE Trans. Ind. Electron.* **2009**, *24*, 1198–1208. [CrossRef]

15. Soto, W.D.; Klein, S.; Beckman, W. Improvement and validation of a model for photovoltaic array performance. *Sol. Energy* **2006**, *80*, 78–88. [CrossRef]

16. Tian, H.; Mancilla-David, F.; Ellis, K.; Muljadi, E.; Jenkis, P. A cell-to-module-to-array detailed model for photovoltaic panels. *Sol. Energy* **2012**, *86*, 2695–2706. [CrossRef]

17. Ding, K.; Zhang, J.; Bian, X.; Xu, J. A simplified model for photovoltaic modules based on improved translation equations. *Sol. Energy* **2014**, *101*, 40–52. [CrossRef]

18. DiPiazza, M.C.; Luna, M.; Petrone, G.; Spagnuolo, G. Translation of the Single-Diode PV Model Parameters Identified by Using Explicit Formulas. *IEEE J. Photovoltaics* **2017**, *7*, 1009–1016.

19. Easwarakhanthan, T.; Bottin, J.; Bouhouch, I.; Boutrit, C. Nonlinear Minimization Algorithm for Determining the Solar Cell Parameters with Microcomputers. *Int. J. Sol. Energy* **1986**, *4*, 1–12. [CrossRef]

20. Ma, T.; Yang, H.; Lu, L. Development of a model to simulate the performance characteristics of crystalline silicon photovoltaic modules/strings/arrays. *Sol. Energy* **2014**, *100*, 31–41. [CrossRef]

21. Balzani, M.; Reatti, A. Neural network based model of a PV array for the optimum performance of PV system. In Proceedings of the Research in Microelectronics and Electronics, Lausanne, Switzerland, 28–28 July 2005.

22. Almonacid, F.; Rus, C.; Hontoria, L.; Munoz, F. Characterisation of PV CIS module by artificial neural networks. A comparative study with other methods. *Renew. Energy* **2010**, *35*, 973–980. [CrossRef]

23. Zagrouba, M.; Sellami, A.; Bouaïcha, M.; Ksouri, M. Identification of PV solar cells and modules parameters using the genetic algorithms: Application to maximum power extraction. *Sol. Energy* **2010**, *84*, 860–866. [CrossRef]

24. Jun, S.J.; Kay-Soon, L. Photovoltaic model identification using particle swarm optimization with inverse barrier constraint. *IEEE Trans. Power Electron.* **2012**, *27*, 3975–3983.

25. Wu, L.; Chen, Z.; Long, C.; Cheng, S.; Lin, P.; Chen, Y.; Chen, H. Parameter extraction of photovoltaic models from measured I-V characteristics curves using a hybrid trust-region reflective algorithm. *Appl. Energy* **2018**, *232*, 36–53. [CrossRef]
26. Nassar-Eddine, I.; Obbadi, A.; Errami, Y.; Fajri, A.E.; Agunaou, M. Parameter estimation of photovoltaic modules using iterative method and the Lambert W function: A comparative study. *Energy Convers. Manag.* **2016**, *119*, 37–48. [CrossRef]
27. Spagnuolo, G. An Interval Arithmetic-based Yield Evaluation in Circuit Tolerance Design. In Proceedings of the IEEE International Symposium on Circuits and Systems. Proceedings, Phoenix-Scottsdale, AZ, USA, 26–29 May 2002.
28. Spagnuolo, G. Worst Case Tolerance Design of Magnetic Devices by Evolutionary Algorithms. *IEEE Trans. Magn.* **2003**, *39*, 2170–2178. [CrossRef]
29. Xu, J.; Wu, Z.; Yu, X.; Hu, Q.; Dou, X. An Interval Arithmetic-Based State Estimation Framework for Power Distribution Networks. *IEEE Trans. Ind. Electron.* **2019**, *66*, 8509–8520. [CrossRef]
30. Petrone, G.; Spagnuolo, G.; Vitelli, M. Analytical model of mismatched photovoltaic fields by means of Lambert W-function. *Sol. Energy Mater. Sol. Cells* **2007**, *91*, 1652–1657. [CrossRef]
31. Accarino, J.; Petrone, G.; Ramos-Paja, C.; Spagnuolo, G. Symbolic Algebra for the Calculation of the Series and Parallel Resistances in PV module model. In Proceedings of the 2013 International Conference on Clean Electrical Power (ICCEP), Alghero, Italy, 11–13 June 2013.
32. IEEE Computer Society; M.S.C. *IEEE Standard for Interval Arithmetic*; IEEE: New York, USA, 2018.
33. Moore, R.E. *Interval Analysis*; Prentice-Hall: Englewood Cliffs, NJ, USA, 1966.
34. Wolf, P.; Benda, V. Identification of PV solar cells and modules parameters by combining statistical and analytical methods. *Sol. Energy* **2013**, *93*, 151–157. [CrossRef]
35. Yordanov, G.H.; Midtgård, O. Physically-consistent parameterization in the modeling of solar photovoltaic devices. In Proceedings of the 2011 IEEE Trondheim PowerTech, Trondheim, Norway, 19–23 June 2011; pp. 1–4. [CrossRef]
36. Jain, A.; Kapoor, A. A new method to determine the diode ideality factor of real solar cell using Lambert W-function. *Sol. Energy Mater. Sol. Cells* **2005**, *85*, 391–396. [CrossRef]
37. Parra, J.S.; Ospina, B.; Mejia, E.F.; Orozco-Gutierrez, M.; Bastidas-Rodríguez, J. Microcontroller Based Low Cost and Modular Architecture for Photovoltaic Array Monitoring. In Proceedings of the 2018 IEEE International Conference on Environment and Electrical Engineering and 2018 IEEE Industrial and Commercial Power Systems Europe (EEEIC/I&CPS Europe), Palermo, Italy, 12–15 June 2018.

Article

Condition Monitoring in Photovoltaic Systems by Semi-Supervised Machine Learning

Lars Maaløe [1,2], Ole Winther [2], Sergiu Spataru [3] and Dezso Sera [4,*

[1] Corti, Copenhagen, 1255 København, Denmark; lm@corti.ai
[2] Applied Mathematics and Computer Science, Technical University of Denmark, 2800 Lyngby, Denmark; olwi@dtu.dk
[3] Department of Energy Technology, Aalborg University, 9100 Aalborg, Denmark; ssp@et.aau.dk
[4] School of Electrical Engineering and Robotics, Queensland University of Technology, Brisbane City, QLD 4000, Australia
* Correspondence: dezso.sera@qut.edu.au

Received: 12 December 2019; Accepted: 20 January 2020; Published: 27 January 2020

Abstract: With the rapid increase in photovoltaic energy production, there is a need for *smart* condition monitoring systems ensuring maximum throughput. Complex methods such as drone inspections are costly and labor intensive; hence, condition monitoring by utilizing sensor data is attractive. In order to recognize meaningful patterns from the sensor data, there is a need for expressive machine learning models. However, supervised machine learning, e.g., regression models, suffer from the cumbersome process of annotating data. By utilizing a recent state-of-the-art semi-supervised machine learning based on probabilistic modeling, we were able to perform condition monitoring in a photovoltaic system with high accuracy and only a small fraction of annotated data. The modeling approach utilizes all the unsupervised data by jointly learning a low-dimensional feature representation and a classification model in an end-to-end fashion. By analysis of the feature representation, new internal condition monitoring states can be detected, proving a practical way of updating the model for better monitoring. We present (i) an analysis that compares the proposed model to corresponding purely supervised approaches, (ii) a study on the semi-supervised capabilities of the model, and (iii) an experiment in which we simulated a real-life condition monitoring system.

Keywords: photovoltaic systems; condition monitoring; fault detection; machine learning; semi-supervised learning

---

1. Introduction

With an ever increasing growth in photovoltaic (PV) energy production, the sheer size of individual power plants is growing at a rapid pace [1]. Building and operating such PV plants has become a viable business in many countries. High PV energy production and maximized yield are fundamental for a profit margin. The challenge is not solely detecting an anomaly in the PV power plant, but also optimizing the operation and maintenance costs once detected [2]. Condition monitoring plays a crucial role, since it is key to identifying the specific system state to ascertain its impact on energy production and ensure minimal maintenance costs; e.g., panel cleaning and replacements, and circuit or diode checks [3]. Another challenge is the size of the PV plants. Minimally, the performance of the strings or arrays needs to be monitored. In a MW range there will be hundreds of PV performance computational streams to monitor in real time or periodically [2].

Many PV plant conditions can result in decreased yield. Amongst the conditions are (i) weather patterns, (ii) PV panel aging, (iii) evolving faults, e.g., diode failure or glass breakage, and (iv) faulty installation of the PV panels [4]. It is quite simple to detect anomalies in energy production; however, it is more complex to find the sources of the anomalies accurately. Furthermore, the cause may be a result of a chain of events for which the causality is very much non-trivial.

Several alternatives for better condition monitoring exist, many of which include quite costly add-ons; e.g., increased amount/accuracy of sensors and infrared inspection [5]. Another complementary approach is traditional statistical analysis of the data [6], but this is resource intensive. A less expensive alternative is a data-driven approach in which supervised machine learning models parameterized by, for example, neural networks, learn from the vast amount of incoming sensor data. These machine learning models have proven efficient in terms of noise resiliency and for finding non-linear correlations within condition monitoring for wind energy [7,8] and PV plants [9–11]. However, there is an inherent problem in assumptions made when applying highly expressive neural networks to the problem of condition monitoring, since they are mostly formulated in a supervised setting. This means that we generally expect a large dataset containing condition-data with adhering labels. Therefore, in order to get started, one must (i) predefine all potential non-overlapping conditions that may happen in a PV plant, (ii) have a vast distribution of annotated data-points for each condition, and (iii) expect no anomalies from the already defined problem. It is quite clear that executing (i) will introduce a constraint on how specific we can be in defining a condition, since many have a tendency to overlap. Task (ii) is also limiting since the data of a PV plant is not directly interpretable by a human. Therefore, one needs to engage in a costly annotation of data-points in order to train the relatively data-hungry neural networks. Finally, (iii) is posing a limit of supervised neural networks, since they are normally not modeled with an uncertainty, resulting in a risk of an overly confident estimate of a severe anomaly [12].

Before proposing a solution to the above, it is important to specify how PV plant condition-data can be defined. In this research the conditions are expressed by the output of sensors, monitoring the PV array current, voltage, in-plane irradiance, external temperature, PV module temperature, and wind speed. The sensor inputs are recorded with a specific temporal resolution. The hypothesis is that, in cohesion, all of these sensor inputs will have unique patterns representing a PV plant condition. We propose a state-of-the-art semi-supervised probabilistic machine learning framework that can capture the unique patterns and cluster them according to their respective similarities. Furthermore, as part of the framework, a supervised classifier, taught with a pre-defined annotation process, categorizes each of these clusters. The probabilistic framework thus models the joint distribution of the condition data and the PV plant state. This should be contrasted to traditional supervised approaches that model the state given the condition data. The big advantage of the model is that it can capture condition data anomalies while also classifying known conditions. In addition to this, the number of annotated data-points needed is very low.

The machine learning framework works by learning a distribution over the PV power plant conditions, and thereby correlates new data points with the learned distribution. In recent years there have been several notable contributions within probabilistic semi-supervised learning methods. Amongst them are [13,14], which utilize the variational auto-encoder framework (VAE) [15,16] for a Bayesian approach to modeling the joint probability between the data and labels. In this paper we utilize the skip deep generative model (SDGM) from [14].

The paper is structured such that we give a background to PV condition monitoring, supervised machine learning for fault detection, and the SDGM. Next we introduce the experimental setup followed by results. We show that SDGM can indeed be used as a machine learning model for condition monitoring, and performs significantly better than its supervised counterparts, even in a fully supervised setting. Finally, we simulate a real-life condition monitoring setup where PV plant conditions are introduced sequentially. In these experiments we show how SDGM is able to detect anomalies, and that retraining the system improves condition monitoring performance.

2. Detection and Identification of PV Power Loss and Failures trough Classification Methods

2.1. PV Failures and Factors Causing Power Loss

There exists a number of external factors that can cause power loss in a PV system, in addition to PV specific degradation modes [4]. These can be roughly categorized into three groups. The first group covers optical losses and degradation, such as soiling, snow, or shading affecting the module surface [17], and discoloration of the encapsulant [4]. These optical power loss factors can be relatively easily detected through visual inspection; however, this is not always feasible for large or hard to reach PV installations. Moreover, detecting them from production measurements can be difficult, since their associated failure patterns in the power measurements are irregular, depending on the size and relative position of the soiling, shading, etc. Detecting such failures is important, since some of them can be remedied relatively easy, through cleaning of the PV panels.

A second category of factors causing power loss in a PV system, is the degradation of the electrical circuit of the PV module. In the most severe cases, these are represented by open-circuit and short-circuit faults within the PV array and associated cabling [4]. But there can also be partial degradation, due to moisture ingress and corrosion of the electrical pathways [18], causing an increased series resistance of the PV array [19]. Such faults are generally difficult to detect through visual inspection, and require thermal IR imaging or electroluminescence to detect. However, they cause more predictable patterns in the production measurements, such as voltage drops proportional to the increase in series resistance. Such failures can cause localized heating and hot-spots, posing a risk of arcing and fire.

The third category corresponds to degradation of the solar cells, which in turn can occur due to a number of stress factors, such as: (i) thermo-mechanical stress, causing solar cell cracks, associated with increased series resistance, shunting, and localized heating [4,19]; (ii) voltage stress, causing potential-induced degradation, primarily associated with a decrease in the cells' shunt resistance, but also corrosion and delamination in the case of some thin film technologies [20]; (iii) diurnal and seasonal variations affecting solar cells with metastable performance behavior, such as certain thin film technologies [21]. Degradation modes in this category are more difficult to detect, and the associated failure patterns in production measurements are more complex. Nonetheless, identifying such failures in their incipient phase is of utmost importance, since they are symptoms of more serious, system-wide problems, such as bad system design, installation practice, or module quality, which should be resolved while the modules and PV system are still in warranty.

The types of power loss factors and degradation modes that can affect PV systems are varied and difficult to formalize. And, only a few of them may affect a PV system within its lifetime, depending mainly on the solar cell technology, panel design and quality, environmental and operational conditions, and installation and maintenance practices.

2.2. Failure Detection through Supervised Classification

Two of the main prerequisites for implementing supervised classification in a condition monitoring system, are: (i) the a priori knowledge of the fault types/classes that will occur/need to be detected in the PV system; and (ii) representative measurement datasets for each of the fault classes, necessary for training the classification model. Once these perquisites are met, and appropriately monitored, production variables are chosen as input, and classifiers are trained for each fault class. Once trained, each classifier will operate continuously, monitoring the production variables, and will be able to discern if the system is in normal operation, or if a specific fault class has occurred.

Many types of supervised classification algorithms exist; e.g., support vector machines (SVM) [22], random forest (RF) [23], and multilabel logistic regression (MLR). These are all very expressive models; however, with the rise of deep learning [24], we have seen a multitude of improvements from models that can capture highly non-linear correlations in the data. The improvements mainly concern areas such as image classification [25] and automatic speech recognition [26]. However, the more *expressive* models also gain traction within renewable energy; e.g., for condition monitoring in wind turbines [8] and as forecasting models for solar irradiance [27]. Defining the deep neural network is not a simple task, due to the vast number of choices that need to be taken in regard to type of architecture, depth, regularization, and much more.

The main challenge in implementing a supervised classification algorithm for detecting faults in a PV system is obtaining the necessary PV production measurement datasets characterizing the different fault classes. Since there are no standardized fault classes and representative datasets, faults of different types and severity can occur throughout the 25+ year expected lifetime of the PV system.

2.3. Proposed Failure Detection through Semi-Supervised Classification

A possible solution is to combine a supervised classification method with a data clustering method that is able to detect anomalous patterns in the monitored PV production data. Next, on-site inspection of the event/fault by maintenance personnel, can help identify the type or *class* of this event/fault. The associated production measurements can then be used to retrain a supervised classifier for the detected event/fault class, such that future instances of the event/fault will be automatically detected and identified by the condition monitoring system, which continuously learns new fault classes as it operates (Figure 1).

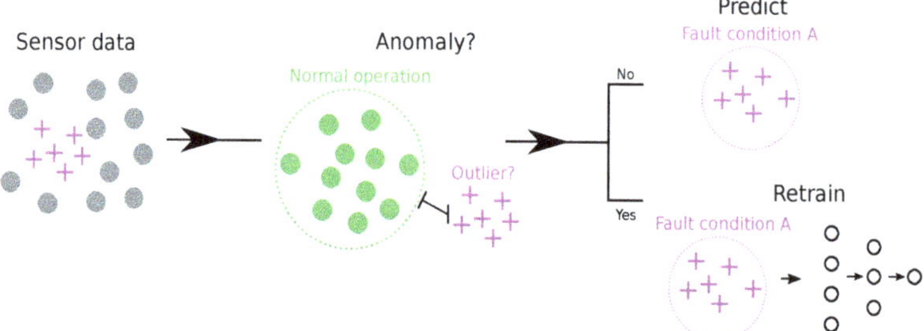

Figure 1. A visualization of the condition monitoring system. The sensor data is propagated through the machine learning framework, and the model detects whether the data point is an outlier. If it is an outlier, the sensor data must be manually inspected, and the machine learning framework retrained. If the incoming sensor data is not an outlier, the framework will predict the state of the condition. If the fault state is detected, as a fault, maintenance will be scheduled accordingly.

We propose to solve the problem for semi-supervised condition monitoring by teaching a feature representation z of the PV condition data x as a continuous conditional probability density function, $p(z|x)$, and the classification task of the PV state y as a discrete conditional probability density function, $p(y|x)$.

In order to teach both models jointly from both labeled and unlabeled data, the two models must be defined such that they share parameters. By applying Bayes theorem we can formulate the problem by:

$$p(z,y|x) = \frac{p(x|z,y)p(z)p(y)}{\int_{z,y} p(x|z,y)p(y)p(z)dz}, \tag{1}$$

where we assume the latent variable feature representation z and state labels y are to be a priori statistically independent, $p(z,y) = p(z)p(y)$. In a scenario with complex input distributions, e.g., sensor input from a PV power plant, the posterior $p(z,y|x)$, becomes intractable. Therefore, we formulate the problem such that we learn an approximation, $q(z,y|x)$, to the posterior through variational inference [28]. SDGM is an example of this probabilistic framework which enables the use of stochastic gradient ascent methods for optimizing the parameters of the generative model, $p_\theta(x,y,z)$, and the variational approximation, $q_\phi(z,y|x)$. θ and ϕ denote the parameters of the generative model and the variational approximation (also denoted inference model) respectively. Both are constructed from deep neural networks (cf. Figure 2). We learn the model parameters by jointly maximizing the objective $\mathcal{L}(x_l,y_l)$ for labeled data x_l,y_l and $\mathcal{U}(x_u)$ for unlabeled data x_u:

$$\mathcal{J} = \sum_{x_l,y_l} \mathcal{L}(x_l,y_l) + \sum_{x_u} \mathcal{U}(x_u). \tag{2}$$

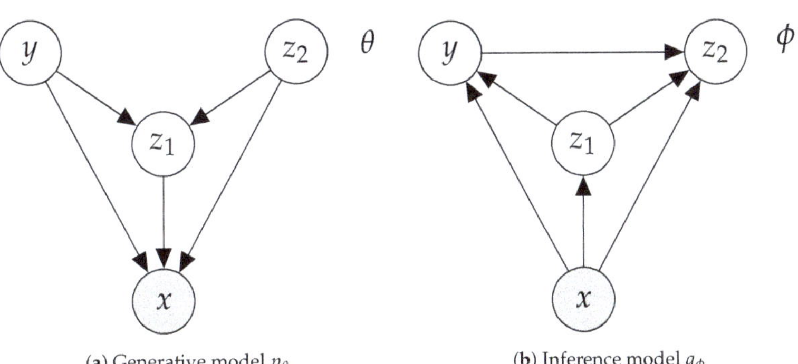

(**a**) Generative model p_θ (**b**) Inference model q_ϕ

Figure 2. The graphical model of the SDGM for semi-supervised learning [14]. The model is defined by two continuous latent variables, z_1 and z_2, a partially observed discrete latent variable y, and a fully observed input x. (**a**) The generative model and (**b**) the inference model, also known as the variational approximation. Each union of incoming edges to a node defines a densely connected deep neural network.

SDGM defines two continuous latent variables, $z = z_1, z_2$, and the discrete partially observed latent variable y [14]. The continuous distributions for the latent variables z are defined as Gaussian distributions and the discrete distribution y is a Categorical distribution. For the labeled data we optimize the parameters, θ,ϕ with respect to a lower bound on the evidence $p(x)$ (ELBO):

$$\log p_\theta(x,y) = \log \int_{z_1} \int_{z_2} p_\theta(x,y,z_1,z_2)dz_2dz_1 \geq \mathbb{E}_{q_\phi(z_1,z_2|x,y)}\left[\log \frac{p_\theta(x,y,z_1,z_2)}{q_\phi(z_1,z_2|x,y)}\right] \equiv \mathcal{F}(x,y), \tag{3}$$

with

$$q_\phi(z_1, z_2 | x, y) = q_\phi(z_1 | x) q_\phi(z_2 | z_1, y, x), \tag{4}$$

$$p_\theta(x, y, z_1, z_2) = p_\theta(x | z_1, z_2, y) p_\theta(z_1 | y, z_2) p_\theta(y) p_\theta(z_2). \tag{5}$$

Since the labeled ELBO does not include the classification error, we add the categorical cross-entropy loss

$$\mathcal{L}(x, y) = \mathcal{F}(x, y) + \alpha \cdot \mathbb{E}_{q_\phi(z_1, z_2 | x, y)} \left[\log q_\theta(y | z_1, x) \right], \tag{6}$$

where α is a constant scaling term defined as a hyper-parameter. Similarly to the labeled loss, we define the unlabeled loss as the unlabeled ELBO:

$$\log p_\theta(x) = \log \int_{z_1} \sum_y \int_{z_2} p(x, y, z_1, z_2) dz_2 dz_1 \geq \mathbb{E}_{q_\phi(z_1, y, z_2 | x)} \left[\log \frac{p_\theta(x, y, z_1, z_2)}{q_\phi(z_1, y, z_2 | x)} \right] \equiv \mathcal{U}(x), \tag{7}$$

where

$$q_\phi(z_1, z_2, y | x) = q_\phi(z_1 | x) q_\phi(y | z_1, x) q_\phi(z_2 | z_1, y, x). \tag{8}$$

In this paper we restrict the experiments to only use densely connected neural networks, but simple extensions to the model include recurrent neural networks and convolutional neural networks that have proven efficient in modeling temporal and spatial information within condition monitoring [8,27]. Besides being among the state-of-the-art within semi-supervised image classification, SDGM posses another intriguing property for condition monitoring, differentiating it from other semi-supervised approaches. Since we are optimizing the ELBO, we can use this as an anomaly measure. Thus, if the value of the ELBO for a specific data-point is far below the value of the unlabeled ELBO, $\mathcal{U}(x)$ that was evaluated during optimization, we can define the data-point as an anomaly.

3. Experimental Application and Tests

In order to validate whether the proposed SDGM can be utilized for condition monitoring, we have recorded a dataset of the sensor data from a small-scale PV plant. During the timespan of the recording, we witnessed 10 different categories that we used as labels. In order to benchmark the machine learning framework, we have defined two comparable supervised machine learning models.

3.1. Field Test Setup and Dataset

To evaluate the progressive learning and fault detection capabilities of the proposed condition monitoring system, we performed measurements and tests on a 0.9 kWp roof-mounted PV string (eight multicrystalline silicon modules). The PV string was connected to a 6 kWp Danfoss TLX Pro string inverter that was continuously monitoring the string current (I), voltage (V), plane-of-array irradiance (G), external temperature (TExt), module temperature (TMod), and windspeed (W), with a one minute sampling time. Since the test PV system is normally not affected by any faults, we created seven power loss events/fault classes by applying different types of shading on the panels, and by connecting different power resistors on series with the PV string, to emulate series resistance type faults. In addition, we also recorded PV production for when the PV system was covered by snow, for a clear sky, and for a cloudy sky day. The ten conditions/fault classes are outlined in Table 1, and will be used as *class labels* for testing the classifiers in the next sections. Another important step in designing a classification model is choosing appropriate input variables. Minimally, PV array current and voltage are monitored in a PV system, and we denote this case as the *simple monitoring* case. Additional monitoring input variables can be the solar irradiance, module temperature, external temperature, and wind speed. These are less commonly monitored in small PV installations, due to the additional costs of the sensors; however, in larger PV plants, these are usually monitored by accurate weather stations. We will denote the case including the ambient conditions as input variables, the *complex monitoring* case.

Table 1. An overview of the PV system dataset used for this research. The dataset comprises 10 categories from approximately 15,000 data samples of a varied representation.

Condition	Description	Samples
PS7	Uniform shading on all lower cells of the modules	10.68%
RS4	50% increase in string series resistance	10.18%
PS50	Partial shading on 50% of a submodule	10.83%
RS8	100% increase in string series resistance	5.11%
PSRS	Combined 50% shading on a submodule with 50% increase in string Rs	10.93%
PS75	Shading on 50% of a submodule + 25% of another submodule	10.60%
C	Cloudy sky day	4.60%
S	Snow on the modules	27.64%
N	Clear sky day	4.67%
IV	Shading on 3/4 of cell area of 6 submodules	4.78%

The categories are skewed in accordance to the weather pattern during the two months; e.g., there is a majority of data points for which there was snow (cf. Table 1). For each learned model, we ran a 5-fold Monte Carlo cross-validation with a random split of 80% for training and 20% for testing. The labeled samples are either sampled uniformly or progressively for each PV system state category.

3.2. Machine Learning Setup

In order to evaluate the proposed machine learning framework, we first define a solid baseline for comparison. Since the SDGM is parameterized by neural networks, we construct a supervised neural network for classification with similar parameterization to $q_\phi(y|z_1, x)$. Furthermore, we also define a simple linear classification model, in order to conclude whether the added complexity from the neural networks is needed for modeling this dataset. The supervised deep neural network for classification is denoted multi-layer perceptron (MLP), and the linear model is referred to as multi-label regression (MLR).

(i) In the first experiment we benchmark SDGM against MLP and MLR in a fully supervised setting; thus, all labels for the entire dataset are given during training. The aim of this experiment is to see whether MLP performs significantly better than MLR and whether SDGM performs approximately equivalently to MLP. We perform this experiment on both the simple and complex monitoring case. (ii) Next, we investigate the semi-supervised performance of the SDGM. In order to do this, we simulate a scenario where only a fraction of data in the PV sensor dataset is given. Since MLP and MLR are supervised models, they are only able to learn from this fraction of labeled data, whereas SDGM can utilize the unlabeled fraction also. The fraction of labeled data is randomly sampled uniformly across categories, such that there is an even representation of each category in the labeled dataset. (iii) Finally, we simulate a real-life PV plant condition monitoring system, in which we assume that each condition is introduced to the power plant sequentially (cf. Figure 1). First, we initialize the dataset with only one labeled data-point from each category, in order to introduce the minimal amount of categorical knowledge in the classifier. Next we introduce 500 labeled samples from the first category in Table 1 and optimize MLP and SDGM. Then we could estimate the ELBO in Equation (7) for the data-points of the categories that are included during training and the ones that are not. We also estimate the accuracy of each classifier in SDGM and the MLP. We expected that the estimate of the ELBO would be significantly lower for the categories that are not included in the dataset as opposed to the ELBO for the categories that are included. This indicates an anomaly. Then, we progressively include another category from Table 1 and perform the same analysis until we have evaluated 6 categories.

The SDGM (For details on experimental implementation and code refer to [14] and the corresponding Github repository.) consists of 2 densely connected deep neural neural networks with parameters θ in the generative model and 3 densely connected deep neural networks with parameters ϕ in the inference model. The neural networks in both the SDGM and MLP contains 2 hidden layers with 50 units in each. We use the ReLU [29] activation function as a non-linearity and ADAM [13] for optimizing the parameters. For the MLP we use a dropout [30] rate of 0.5 and for the MLR we use L2 regularization. Model training is stopped upon saturation of the validation error. The α constant is defined as in [14]. During optimization of the SDGM we utilize the warm-up introduced in [31,32].

4. Results

We performed three experiments, introduced above. In the first experiment we benchmarked the SDGM against the MLP and MLR in a fully-supervised setting. Next we evaluated the semi-supervised power of the SDGM. Finally, we simulated a real-life condition monitoring system.

4.1. Supervised Condition Monitoring Accuracy

Table 2 presents the baseline results of MLR, MLP, and SDGM in a fully supervised learning setup. By utilizing more sensor attributes (complex versus simple), the performance increases well over 10% across all models. This proves that the additional sensor inputs (G, TExt, TMod, and W) are very useful for condition monitoring. When comparing the non-linear MLP to the linear MLR we also achieve a significant improvement in performance, indicating that the input data is not linearly separable, and that the added complexity of the neural networks is worthwhile.

Table 2. Fully supervised baselines of MLR, MLP, and SDGM with the *simple* sensor input, {I, V}, and the *complex* input, {I, V, G, TExt, TMod, and W}.

	Accuracy I, V	Accuracy I, V, G, TExt TMod, W
MLR	51.62%	77.33%
MLP	77.81%	89.11%
SDGM	**79.06%**	**92.47%**

The most surprising finding was that the SDGM performs significantly better than MLP. We believe that this is due to the fact that SDGM also learns a latent clustering of the data that is correlated with the PV state.

Thereby, the model can discriminate between the labels and the cluster representations, meaning that it can put less emphasis on labeled information that does not seem to correlate with the distribution. Hence, if a small fraction of faulty labels exist in the training dataset, SDGM is able to ignore this information and thereby achieve better generalization towards the validation dataset.

Figure 3 shows how the wrongly classified examples from the MLR and MLP are quite similar. The highest misclassification rate lies between cloudy and snowy weather, {C, S}. Other misclassification rates mainly lie between {RS8, PS50}, {N, RS4}, {RS8, IV}, and {N, RS4}. When we compare the results of MLR and MLP to the SDGM (cf. Figure 4a) we can read from the confusion matrix that the SDGM manages to learn the difference between cloudy and snowy, {C, S}. Furthermore the remainder of the most prominent misclassification rates are significantly decreased. In order to analyze what is learned in the latent variables of SDGM, we plot the first two principal components from a principal component analysis (PCA) (cf. Figure 4b). The visualization of the latent space shows clear discrimination between categories. Furthermore, we can also see that the data lies on manifolds resembling the movement of the sun.

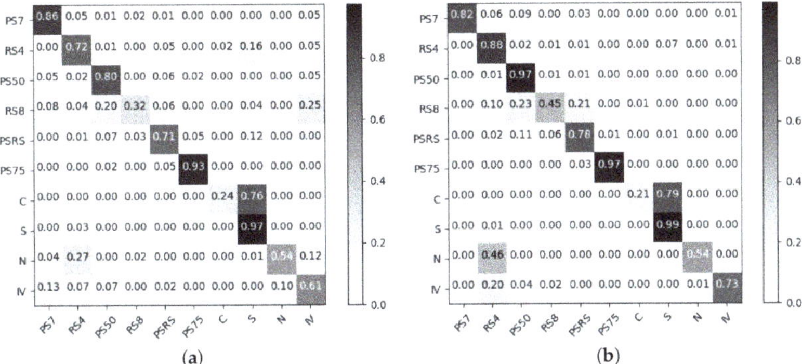

(a) (b)

Figure 3. Normalized confusion matrices for (**a**) MLR and (**b**) MLP trained on the fully labeled complex dataset. The x-axis denotes the predicted labels and the y-axis the true labels.

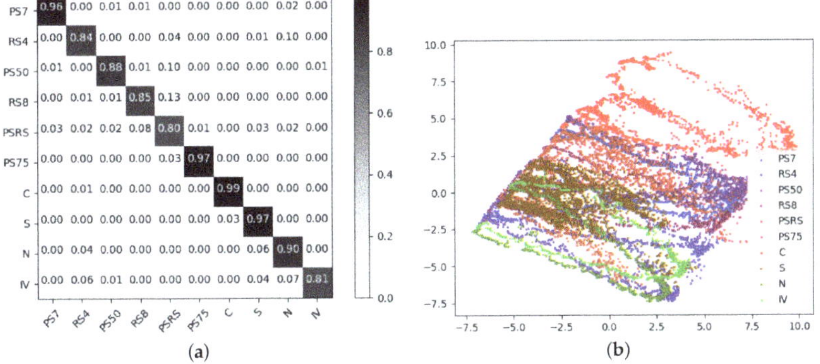

(a) (b)

Figure 4. (**a**) Normalized confusion matrices for SDGM trained on the fully labeled *complex* dataset. (**b**) PCA (principal components 1 and 2) on the latent space.

4.2. Semi-Supervised Condition Monitoring Accuracy

In order to evaluate the semi-supervised performance of SDGM, we define eight datasets with different fractions of labeled data that are randomly subsampled across the categories in Table 1 for each of the trained models, {100, 300, ..., 1500}. Figure 5 shows SDGM's significant increase in performance by utilizing the information in the unlabeled data. For the simple dataset, with {I, V} as input, we see that the supervised models, MLR and MLP, achieve an accuracy of 35%–45% by learning from 100 labeled data-points, whereas the SDGM achieves 55%–60%. As expected, the relative improvement from using SDGM stays significant when introducing more labels. Similarly to the supervised analysis above, all models achieve a significant improvement when adding more sensor inputs, {I, V, G, TExt, TMod, W}. When comparing the results of the semi-supervised SDGM with the supervised SDGM, we see that the models trained on 1500 labeled data points actually exceed the performance of the fully-supervised model, 93.12% compared to 92.47%.

Again, the reason for this may be that with fewer labeled examples, SDGM put a larger emphasis on the unlabeled data, and thereby it was not as prone to faulty annotations. In Figure 6 we visualize the latent representations by PCA for the SDGM trained with 100 labeled data-points on the simple and complex input. It is clear that the model trained on the complex is better at discriminating between the categories than the model trained on the simple input. Furthermore, when comparing Figure 4b with Figure 6b we see clear indications that the increase in labels results in better discrimination between condition states.

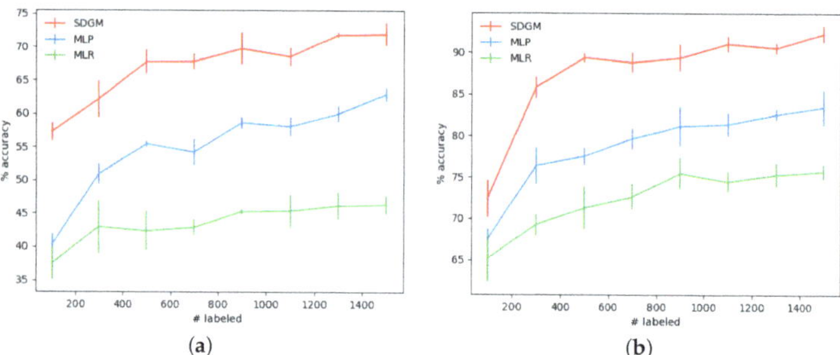

Figure 5. Comparison between the supervised MLP, MLR, and the semi-supervised SDGM trained on an increasing amount of randomly sampled and evenly distributed labeled data points. For each number of labeled data points, we trained 10 different models, since a large variance between the quality of the subsampled labeled data points may have existed. (**a**) The accuracy with one standard deviation for models trained on the simple input distribution {I, V}, and (**b**) the accuracy for models trained on the *complex* input distribution, {I, V, G, TExt, TMod, W}.

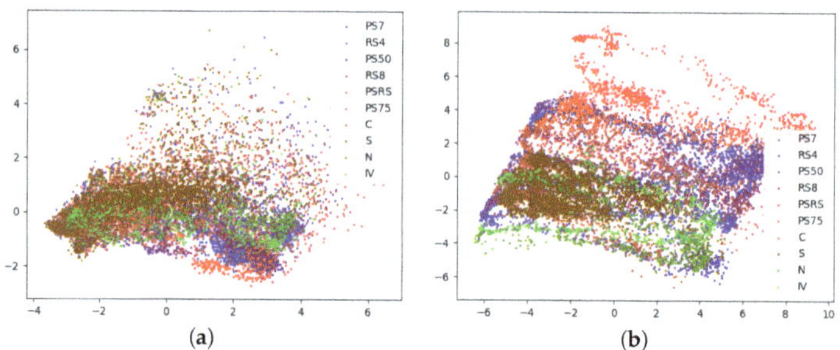

Figure 6. PCA (principal component 1 and 2) visualization of the latent space for SDGMs trained a dataset with only 100 labeled samples. (**a**) The latent space for a model trained on the simple dataset, {I, V} and (**b**) for {I, V, G, TExt, TMod, W} as input.

4.3. Adding PV Conditions Progressively

In a PV system it is highly unlikely that a dataset will consist of an evenly distributed labeled dataset from all categories. To test whether the SDGM is able to perform anomaly detection on the data, we set up an experiment where we began by learning a model on 500 randomly sampled labeled data points and only one labeled data point for each of the remaining categories. Then, we progressively taught new models with a dataset to which we added 500 labels for the next category. We continued this procedure until the 6th category.

Figure 7a presents the results of a SDGM and MLP taught up to six categories. As expected, the accuracy for all categories increases when more categories are added to the dataset. Again, it is clear that the SDGM is able to utilize the information from the unlabeled examples and the very sparse information from the other categories to significantly outperform the MLP. In Figure 7b, we visualize the level of certainty and ELBO (cf. Equation (7)), and can easily discriminate the categories included during training from the categories that are not included. So for a model trained on only {PS7} data, it is easy to detect {RS4, PS50, RS8, PSRS, PS75, C, S, N, IV} conditions as anomalous, and for a model trained on {PS7,RS4} it is easy to detect {PS50, RS8, PSRS, PS75, C, S, N, IV} as anomalous. In order to state whether a PV plant condition is an anomaly, the operator needs to define a threshold value. In this experiment a suitable threshold could be that PV plant conditions with an ELBO below -60 nats is considered an anomaly. Upon realization of an anomaly, the PV plant operator will initiate a brief annotation process and retrain the SDGM framework, so that the new states are within the known operational condition.

Figure 8 presents the classification errors for the MLP and SDGM when only taught on 100 labeled data points. Since the SDGM is able to utilize the information of the unlabeled data points it is also able to classify much better across PV plant categories.

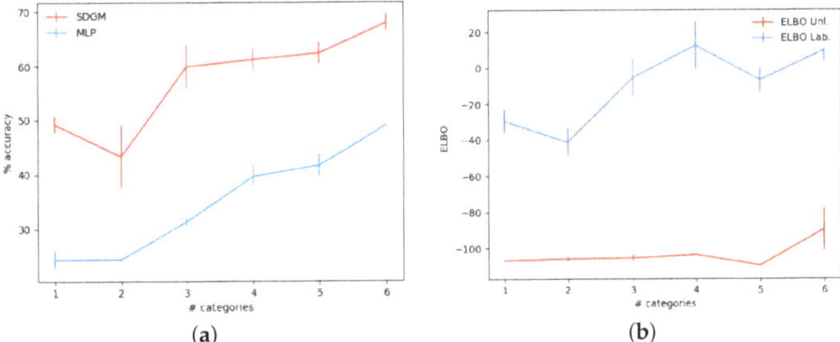

Figure 7. SDGMs and MLPs were trained with datasets from which we randomly subsampled a single data-point from each category and then progressively added 500 randomly labeled data points for each category, and we trained a new MLP and SDGM for each progression. (**a**) The accuracy of the classifiers for the SDGM and MLP. (**b**) The ELBO for the data categories included during training (ELBO Lab.) The data categories that are not included during training (ELBO Unl.). The categories that were progressively added followed the order of Table 1; i.e., first *{PS7}*, and next *{PS7, RS4}*, until reaching *{PS7, RS4, PS50, RS8, PSRS, PS75}*.

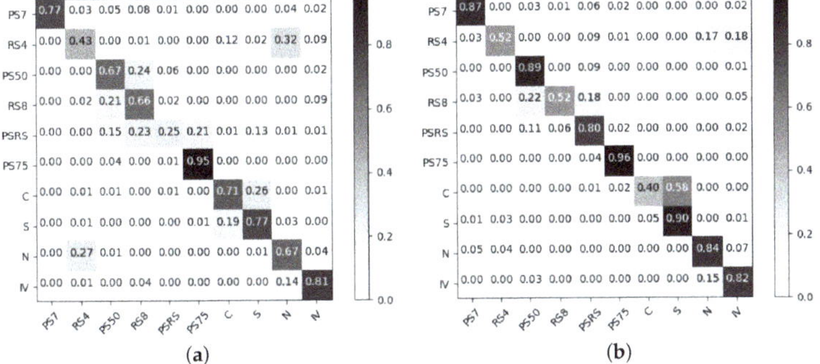

Figure 8. Normalized confusion matrices for (**a**) MLP and (**b**) SDGM trained on 100 randomly sampled and evenly distributed labeled data points. The x-axis shows the predicted labels, and the y-axis the true labels.

5. Conclusions

In this research we have proposed a novel machine learning framework to perform PV condition monitoring that simultaneously learns classification and anomaly detection models. We have shown that the proposed semi-supervised framework is able to improve over a fully supervised framework when given a full set of labeled data points (fully supervised learning) and when only given a fraction of labeled data points (semi-supervised learning). We have also shown that the framework is able to identify previously unknown fault types by performing anomaly detection, and how it can be easily retrained in order to capture these PV states. This approach can significantly improve the throughput of energy production and lower the maintenance cost of PV power plants. We have shown that the approach is easy to train on a rather simple dataset and that it is easily interpretable by evaluating the classification results, the latent representations, and the lower bound of the marginal log-likelihood.

The main limitation of this research lies in the dataset used. Due to the representation and the amount of samples, it does not resemble the vast amount of data one could acquire from a large-scale PV power plant. However, deep neural networks have a tendency to improve when introduced to more data, meaning that we can hypothesize that the results would only improve. In this regard, an interesting direction for future research would be to investigate the possibility for transfer learning between PV power plant configurations, so that one could seamlessly deploy the framework taught on one PV plant to another.

Author Contributions: Conceptualization, L.M. and S.S.; methodology, L.M., O.W., S.S., and D.S; software, L.M.; validation, L.M. and S.S.; formal analysis, L.M. and S.S.; investigation, L.M. and S.S.; resources, L.M., S.S., and D.S.; data curation, S.S. and D.S.; writing—original draft preparation, S.S. and L.M.; writing—review and editing, S.S., D.S., O.W., and L.M.; visualization, L.M.; supervision, D.S. and O.W.; project administration, D.S. and O.W.; funding acquisition, O.W. and D.S. All authors have read and agreed to the published version of the manuscript.

Funding: This research was funded by Danish National Advanced Technology Foundation, grant number HTF 102-2013-3.

Acknowledgments: The research was supported by the NVIDIA Corporation with the donation of TITAN X GPUs.

Conflicts of Interest: The authors declare no conflicts of interest.

References

1. IEA PVPS. *Trends 2017 in Phtovoltaic Applications—Survey Report of Selected IEA Countries between 1992 and 2016*; International Energy Agency: Paris, France, 2018.
2. Mütter, G.; Krametz, T.; Steirer, P. Experiences with a performance package for multi-MW PV plants based on computations on top of monitoring. In Proceedings of the 31st European Photovoltaic Solar Energy Conference and Exhibition WIP, Hamburg, Germany, 14–18 September 2015; pp. 1675–1678.
3. Woyte, A.; Richter, M.; Moser, D.; Reich, N.; Green, M.; Mau, S.; Beyer, H.G. *Analytical Monitoring of Grid-Connected Photovoltaic Systems*; International Energy Agency: Paris, France, 2014.
4. Köntges, M.; Kurtz, S.; Packard, C.; Jahn, U.; Berger, K.A.; Kato, K.; Friesen, T.; Liu, H.; Van Iseghem, M. *Review of Failures of Photovoltaic Modules*; International Energy Agency: Paris, France, 2014.
5. Buerhop-Lutz, C.; Scheuerpflug, H.; Pickel, T.; Camus, C.; Hauch, J.; Brabec, C. IR-Imaging a Tracked PV-Plant Using an Unmanned Aerial Vehicle. In Proceedings of the 32nd European Photovoltaic Solar Energy Conference and Exhibition WIP, 2016, Munich, Germany, 20–24 June 2016; pp. 2016–2020.
6. Vergura, S.; Acciani, G.; Amoruso, V.; Patrono, G.E.; Vacca, F. Descriptive and Inferential Statistics for Supervising and Monitoring the Operation of PV Plants. *IEEE Trans. Ind. Electron.* **2009**, *56*, 4456–4464. [CrossRef]
7. Bach-Andersen, M. A Diagnostic and Predictive Framework for Wind Turbine Drive Train Monitoring. Ph.D. Thesis, Technical University of Denmark, Lyngby, Denmark, 2017.
8. Bach-Andersen, M.; Rømer-Odgaard, B.; Winther, O. Deep learning for automated drivetrain fault detection. *Wind Energy* **2017**, *21*, 29–41. [CrossRef]

9.   Silvestre, S.; Chouder, A.; Karatepe, E. Automatic fault detection in grid connected PV systems. *Sol. Energy* **2013**, *94*, 119–127. [CrossRef]

10. Jiang, L.L.; Maskell, D.L. Automatic fault detection and diagnosis for photovoltaic systems using combined artificial neural network and analytical based methods. In Proceedings of the IEEE International Joint Conference on Neural Networks, Killarney, Ireland, 12–17 July 2015.

11. Ali, M.H.; Rabhi, A.; El Hajjaji, A.; Tina, G.M. Real Time Fault Detection in Photovoltaic Systems. *Procedia Energy* **2017**, *111*, 914–923. [CrossRef]

12. Gal, Y. Uncertainty in Deep Learning. Ph.D. Thesis, University of Cambridge, Cambridge, UK, 2016.

13. Kingma, D.P.; Rezende, D.J.; Mohamed, S.; Welling, M. Semi-Supervised Learning with Deep Generative Models. In Proceedings of the International Conference on Machine Learning, Beijing, China, 21–26 June 2014; pp. 3581–3589.

14. Maaløe, L.; Sønderby, C.K.; Sønderby, S.K.; Winther, O. Auxiliary Deep Generative Models. In Proceedings of the International Conference of Machine Learning, New York, NY, USA, 19–24 June 2016.

15. Kingma, D.P.; Welling, M. Auto-Encoding Variational Bayes. In Proceedings of the International Conference on Learning Representations, Scottsdale, AZ, USA, 2–4 May 2013.

16. Rezende, D.J.; Mohamed, S.; Wierstra, D. Stochastic Backpropagation and Approximate Inference in Deep Generative Models. *arXiv* **2014**, arXiv:1401.4082.

17. Laukamp, H.; Schoen, T.; Ruoss, D. *Reliability Study of Grid Connected PV Systems, Field Experience and Recommended Design Practice*; International Energy Agency: Paris, France, 2002.

18. Yang, B.B.; Sorensen, N.R.; Burton, P.D.; Taylor, J.M.; Kilgo, A.C.; Robinson, D.G.; Granata, J.E. Reliability model development for photovoltaic connector lifetime prediction capabilities. In Proceedings of the 39th IEEE Photovoltaic Specialists Conference (PVSC), Tampa, FL, USA, 16–21 June 2013; pp. 139–144.

19. King, D.L.; Quintana, M.A.; Kratochvil, J.A.; Ellibee, D.E.; Hansen, B.R. Photovoltaic module performance and durability following long-term field exposure. *Prog. Photovolt. Res. Appl.* **2000**, *8*, 241–256. [CrossRef]

20. Luo, W.; Khoo, Y.S.; Hacke, P.; Naumann, V.; Lausch, D.; Harvey, S.P.; Singh, J.P.; Chai, J.; Wang, Y.; Aberle, A.G.; et al. Potential-induced degradation in photovoltaic modules: A critical review. *Energy Environ. Sci.* **2017**, *10*, 43–68. [CrossRef]

21. Silverman, T.; Jahn, U.; Friesen, G.; Pravettoni, M.; Apolloni, M.; Louwen, A.; Schweiger, M.; Belluardo, G. *Characterisation of Performance of Thin-film Photovoltaic Technologies*; International Energy Agency: Paris, France, 2014.

22. Cortes, C.; Vapnik, V. Support-Vector Networks. *Mach. Learn.* **1995**, *20*, 273–297. [CrossRef]

23. Breiman, L. Random Forests. *Mach. Learn.* **2001**, *45*, 5–32. [CrossRef]

24. Hinton, G.E.; Osindero, S.; Teh, Y.W. A Fast Learning Algorithm for Deep Belief Nets. *Neural Comput.* **2006**, *18*, 1527–1554. [CrossRef] [PubMed]

25. He, K.; Zhang, X.; Ren, S.; Sun, J. Deep Residual Learning for Image Recognition. In Proceedings of the IEEE Conference on Computer Vision and Pattern Recognition, Las Vegas, NV, USA, 27–30 June 2016; pp. 770–778.

26. Xiong, W.; Wu, L.; Alleva, F.; Droppo, J.; Huang, X.; Stolcke, A. The Microsoft 2017 Conversational Speech Recognition System. *arXiv* **2017**, arXiv:1708.06073.

27. Alzahrani, A.; Shamsi, P.; Dagli, C.; Ferdowsi, M. Solar Irradiance Forecasting Using Deep Neural Networks. *Procedia Comput. Sci.* **2017**, *114*, 304–313. [CrossRef]

28. Jordan, M.I.; Ghahramani, Z.; Jaakkola, T.S.; Saul, L.K. An Introduction to Variational Methods for Graphical Models. *Mach. Learn.* **1999**, *37*, 183–233. [CrossRef]

29. Glorot, X.; Bordes, A.; Bengio, Y. Deep Sparse Rectifier Neural Networks. In Proceedings of the 14th International Conference on Artificial Intelligence and Statistics (AISTATS), Ft. Lauderdale, FL, USA, 11–13 April 2011; Volume 15, pp. 315–323.

30. Srivastava, N.; Hinton, G.; Krizhevsky, A.; Sutskever, I.; Salakhutdinov, R. Dropout: A Simple Way to Prevent Neural Networks from Overfitting. *JMLR* **2014**, *15*, 1929–1958.

31. Sønderby, C.K.; Raiko, T.; Maaløe, L.; Sønderby, S.K.; Winther, O. Ladder Variational Autoencoders. In *Advances in Neural Information Processing Systems*; Neural Information Processing Systems (NIPS): Barcelona, Spain, 2016.
32. Bowman, S.; Vilnis, L.; Vinyals, O.; Dai, A.; Jozefowicz, R.; Bengio, S. Generating sentences from a continuous space. *arXiv* **2015**, arXiv:1511.06349.

Article

Practical Implementation of the Backstepping Sliding Mode Controller MPPT for a PV-Storage Application

Marwen Bjaoui [1,*]**, Brahim Khiari** [1]**, Ridha Benadli** [1]**, Mouad Memni** [1] **and Anis Sellami** [2]

[1]   LANSER Laboratory/CRTEn B.P.95 Hammam-Lif, Tunis 2050, Tunisia; brahim.khiari@crten.rnrt.tn (B.K.); ridhabenadly@gmail.com (R.B.); memnimouad@hotmail.fr (M.M.)

[2]   Research unit: LISIER, National Higher Engineering School of Tunis, Tunis 2050, Tunisia; anis.sellami@esstt.rnu.tn

*   Correspondence: bjaouimarwen@yahoo.fr; Tel.: +216-28770-446

Received: 26 June 2019; Accepted: 17 July 2019; Published: 16 September 2019

Abstract: This study presents a design and an implementation of a robust Maximum Power Point Tracking (MPPT) for a stand-alone photovoltaic (PV) system with battery storage. A new control scheme is applied for the boost converter based on the combination of the adaptive perturb and observe fuzzy logic controller (P&O-FLC) MPPT technique and the backstepping sliding mode control (BS-SMC) approach. The MPPT controller design was used to accurately track the PV operating point to its maximum power point (MPP) under changing climatic conditions. The presented MPPT based on the P&O-FLC technique generates the reference PV voltage and then a cascade control loop type, based on the BS-SMC approach is used. The aims of this approach are applied to regulate the inductor current and then the PV voltage to its reference values. In order to reduce system costs and complexity, a high gain observer (HGO) was designed, based on the model of the PV system, to estimate online the real value of the boost converter's inductor current. The performance and the robustness of the BS-SMC approach are evaluated using a comparative simulation with a conventional proportional integral (PI) controller implemented in the MATLAB/Simulink environment. The obtained results demonstrate that the proposed approach not only provides a near-perfect tracking performance (dynamic response, overshoot, steady-state error), but also offers greater robustness and stability than the conventional PI controller. Experimental results fitted with dSPACE software reveal that the PV module could reach the MPP and achieve the performance and robustness of the designed BS-SMC MPPT controller.

Keywords: photovoltaic system; maximum power point tracking; backstepping sliding mode control; high gain observer; stability analysis

---

1. Introduction

Most of our electricity needs are met by non-renewable resources which are depleting at a rapid rate. The increasing population and growing needs of energy sources present a motive to look for potential alternatives. In this regard, the production of photovoltaic (PV) energy has drawn a tremendous amount of interest. However, PV energy is still considered expensive and reducing the cost of PV systems has become a main topic of extensive research. To solve the problems above, maximising PV output power can be approached via power electronics [1,2]. The use of MPPT controller for a PV application is crucial to increasing the efficiency and the performance of a PV system [3]. The most used batteries are the lead–acid type, due to their significant autonomy and their reliable and low-cost technology. These rechargeable electrochemical devices are widely employed in many applications such as PV storage systems [1,4,5]. Several MPPT methods have been attempted to track the MPP in PV systems such as perturb and observe (P&O) [6,7], the incremental conductance (INC) method [8–10], the neural network controller (NNC) [11] and the fuzzy logic controller (FLC) [12–14]. The control parameter of the P&O technique is perturbed due to a small variation of the step size. The direction of step size caused by this algorithm is varied due to the measurement of the output power of the PV array. The disturbance of the system depends on the increase or the decrease of power [9]. The increment conductance (IC) technique is based on determining the operating point of the PV module. This method tries to raise the operating point of the PV generator until reaching the MPP. It enables a search of the MPP according to the equality of the conductance and its increment [9,10].

The most commonly employed methods in the literature are the P&O [6] and IC [9], due to the ease of both their understanding and implementation. However, these methods are not efficient during the rapid changing of climatic conditions. Furthermore, even in stable climatic conditions, they produce oscillations around the MPP and they are totally dependent on solar irradiation. In fact, the performance of these methods decreases with the decrease of solar irradiation [9].

The FLC provides the best performance compared to conventional P&O and IC techniques. However, the limitation of this technique comes from its non-achievement of sufficient accuracy of the operating point of the PV generator for the MPP. The step duty cycle changes direction according to the direction change of the adjusted power [15]. Their inputs and outputs depend entirely on the information about the system model to be studied by the designer [16].

Other techniques have been designed such as the MPPT, which is based on the dedicated sliding mode controller for PV storage systems [17,18]. This approach is of great importance given its several advantages such as stability, robustness against the parameter variation, fast dynamic response and the simplicity of implementation [19,20]. However, the SMC-MPPT approach, when applied to the dc–dc converter, has certain drawbacks, including the variability of the operating frequency in the output of the control (chattering phenomenon) [17,21–24]. This study [25] presents an experimental validation of a new SMC for a two-level voltage source inverter for a grid connected PV system. A combination of a traditional MPPT P&O technique and an SMC has been developed in [26]. Similarly, in [27], the authors suggest a backstepping sliding mode control (BS-SMC) scheme to improve the PV system. The MPP seeking method is employed to estimate the reference PV voltage. Then a cascade control loop with a BS-SMC controller aims to regulate the PV voltage to its reference values in order to monitor exactly the PV operating point under variations of the atmospheric conditions.

Moreover, the Hall Effect current sensor is used to measure the inductor current via the Hall Effect, to generate a voltage which is exactly proportional to the current to be measured or visualized. Due to the high sensitivity of this type of sensor to external or parasitic magnetic fields, the measurement of the inductor current can be erroneous. Thus, the performance of the MPPT controller could be reduced [21,28]. Conventional MPPT techniques use Hall Effect sensors and include additional circuitry such as signal conditioning buffers, filters, and amplifier circuits. This increases the cost and complexity and affects the performance system. Unfortunately, once the sensor is damaged, the operation of a photovoltaic generator will be interrupted.

Herein, based on the motivation above, we propose an MPPT based on BS-SMC to enhance the performance and the robustness of the PV storage system. The BS-SMC MPPT method is the combination between the backstepping method and sliding mode. The aim of this approach is to force the system state to achieve the MPP with a high tracking performance and stability. The convergence of the dynamics of the system around the sliding surface depends on two criteria which have been already proposed by Utkin [29] and proved by the Lyapunov function. This method is considered as a potential approach in various applications due to its robustness, its easy implementation and its ability to reject disturbances. This method is presented in reference [30], to control the attitude and position of a quadrotor unmanned aerial vehicle (UAV). The obtained results suggest its relevant performance and robustness, urging its use for the control of the dc–dc boost converter in PV storage system. However, in [27], the authors combined an extremum seeking control method to reach the PV reference voltage with the BS-SMC approach. This method shows these limits underlined by the curve of the sliding surface during variations in climatic conditions. It can be seen in this curve large-amplitude chattering phenomena. In another study [31], an MPPT controller was proposed, based on the combination of the regression plane method and a backstepping controller with integral action (IBS) to control a dc–dc buck boost converter. This technique does not take into account the faults or malfunctioning of the PV module. In addition, IBS provides a minimum of error in the stable state, without satisfying the feature of robustness.

The main goal of this study is to design a new switching function based on Lyapunov stability, to overcome the drawbacks associated with control time and reduce the cost of the PV system. In this context, there are many approaches to mitigating the disadvantages of chattering in SMC, such as using a regular approximation of the switching element or using a higher order sliding mode control (HOSMC) strategy [32].

However, using a continuous approximation affects system performance and requires finite time convergence sliding mode control. In HOSMC, it is generally difficult to estimate the high-order state derivatives and it still presents chattering in the presence of parasitic dynamics. In this study, using a sliding surface including a time function, large-amplitude chattering phenomena are attenuated and thus, robustness is ensured.

The designed MPPT controller is developed to a PV system, including a PV module, a dc–dc boost converter and a battery load. The principle of the studied control scheme contains two cascade control loops. The outer control loop based on the P&O-FLC is used to estimate the real-time of the reference voltage, which corresponds to the maximum power. The inner control loop regulates the PV voltage to reach the value of the reference voltage estimated by the outer control loop. In the absence of the inductor current sensor, the BS-SMC approach is used to track the MPP under solar irradiation and temperature variation, while the estimation of the inductor current of the dc–dc boost converter is carried out by an HGO, as shown in Figure 1. On the other hand, the estimated current is used in the input of the BS-SMC.

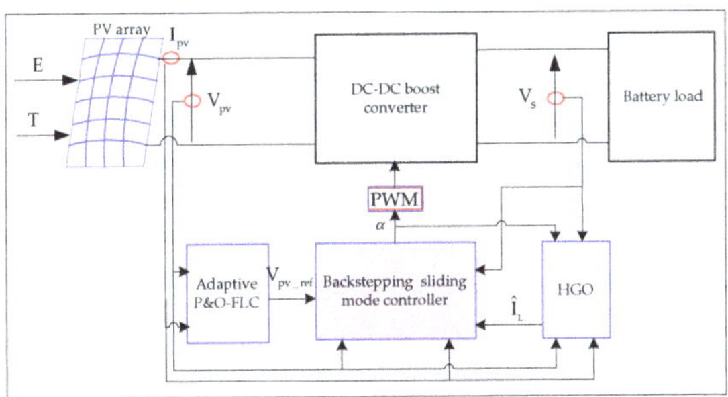

Figure 1. Control scheme of the PV storage system.

After a general introduction, this article first details the modelling of the PV storage system, the design of the BS-SMC controller and the HGO, then the stability analysis. Secondly, we are interested in comparing the simulation results of the proposed BS-SMC controller with the conventional PI controller. The experimental results are illustrated, explained and discussed in detail in the third section. Finally, this paper is completed by a conclusion and perspectives.

2. Modelling of the PV System

Figure 2 shows the proposed configuration of the photovoltaic (PV) storage system which consists of a PV module, a dc–dc boost converter and a battery load.

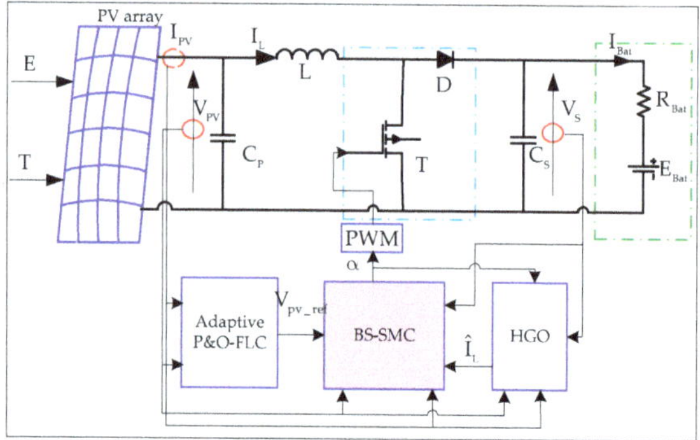

Figure 2. Diagram of the PV storage system.

2.1. Modelling of PV Module

Figure 3 illustrates the equivalent electrical circuit model of the PV panel [1,10,18,33].

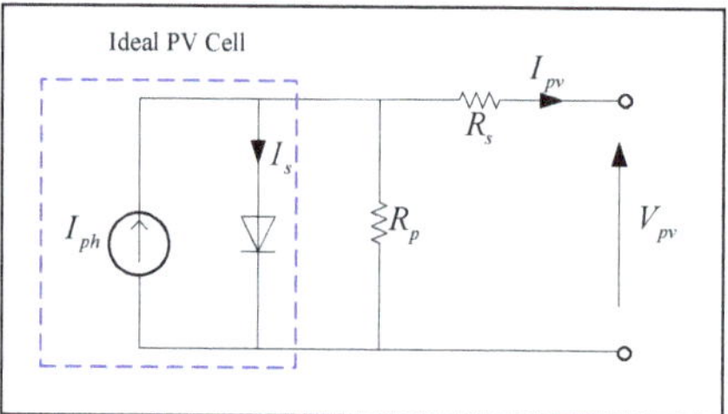

Figure 3. Equivalent circuit model of the PV cell.

The expression of the solar PV cell terminal current as a function of the photo-current, the diode current and the shunt current, can be expressed by:

$$I_{PV} = I_{PH} - I_S - I_{SH} \tag{1}$$

where I_{PH} is the photocurrent of the cell, (A). I_S is the saturation current of the P–N junction and I_{SH} is the current through the parallel resistor R_{SH}.

The output current of PV array can be given by:

$$I_{PV} = N_P I_{PV} - N_P I_S \left(\exp\left(\frac{V_{PV} + R_S I_{PV}}{n_s k T N_S} \right) - 1 \right) - N_P q \left(\frac{V_{PV} + R_S I_{PV}}{N_S R_{SH}} \right) \tag{2}$$

where I_s is cell reverse saturation current; q is the electron charge $\left(q = 1.602 \times 10^{-19} C \right)$; k is the Boltzman constant $\left(k = 1.38 \times 10^{-23} J/K \right)$; n is ideality factor solar cell; V_{PV} is the output voltage; N_P is the number of PV cells connected in parallel; N_S is the number of PV cells connected in series; R_S and R_{SH} are the series and shunt resistors of the PV array, respectively.

The cell reverse saturation current can be determined by:

$$I_S = I_{S0} \left(\frac{T}{T_R} \right)^3 \exp\left(\frac{q E_G}{ka} \left(\frac{1}{T_R} - \frac{1}{T} \right) \right) \tag{3}$$

where $E_G = 1.1(eV)$ is the band gap energy of the semi conductor used in the cell and $T_R(K)$ is the cell reference temperature.

The reverse saturation current I_{S0} at T can be calculated by:

$$I_{S0} = \frac{I_{SC0}}{\exp\left(\frac{qV_{OC}}{N_S aKT}\right) - 1} \tag{4}$$

where $I_{SC0}(A)$ is the cell short-circuit at reference temperature and solar irradiance and the open circuit voltage is $V_{OC}(V)$.

The photocurrent $I_{PH}(A)$ is related to the solar irradiation and temperature; its expression is obtained by the following equation:

$$I_{PH} = \frac{E}{1000}(I_{SC} + k_i(T - T_C)) \tag{5}$$

where $k_i(A/K)$ is the short circuit current temperature coefficient and $E(W/m^2)$ is the solar irradiance.

The monocrystalline Solo Line LX-100M model has been chosen in this paper. The specification parameters of the PV module are presented in Table 1.

Table 1. Parameters of the Solo Line LX-100M PV panel under standard test conditions (STC).

Parameter	Name	Value
P_{MAX}	Maximum Power	100 Wp
V_{mp}	Voltage at maximum power	18.7 V
I_{mp}	Current at maximum power	5.39 A
V_{OC}	Open circuit voltage	21.6 V
I_{SC}	Short circuit current	5.87 A
k_i	Temperature coefficient of I_{SC}	1.73 mA/°C
n_s	Number of cells per module	60

The characteristic I_V and P_V curves of the PV module for different values of solar irradiation E and temperatures T are illustrated in Figure 4a,b, respectively.

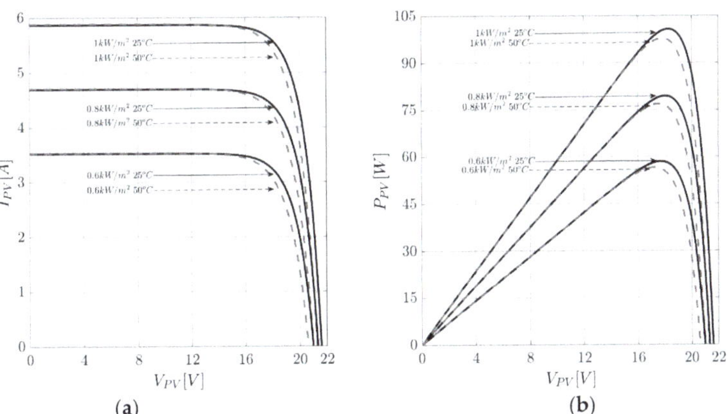

Figure 4. Characteristic curves of PV array obtained under weather conditions: (a) I_V characteristic curves; (b) P_V characteristic curves.

Figure 4a,b depict the general appearance of the electrical characteristics of a PV generator for different values of temperature and solar irradiation. It can be noted that with the increasing temperature, the generated current increases slightly. Inversely, the open circuit voltage decreases considerably. It can be observed that the variation of the solar irradiation greatly affects the short circuit, with low impact on the open circuit voltage. Consequently, the variation of MPP depends on the solar irradiation.

2.2. Modelling of the DC–DC Boost Converter

The block diagram of the dc–dc boost converter is illustrated in Figure 5. It is used to adjust the PV voltage V_{PV} to its reference value corresponding to the MPP.

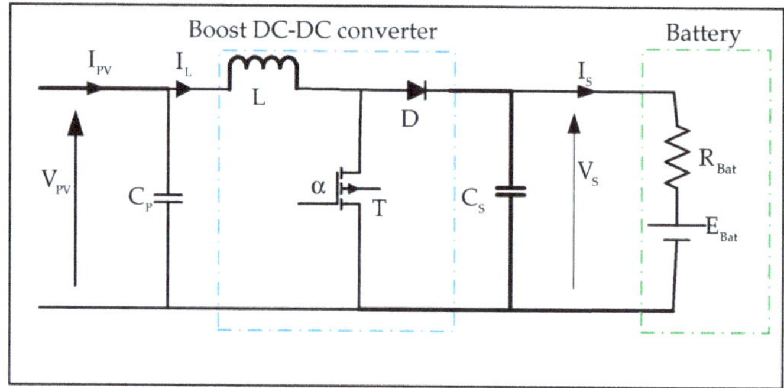

Figure 5. Circuit diagram of dc-dc boost converter.

The system dynamics are described by the following equations [1,9,17,22,33–35]:

$$\begin{cases} \frac{dV_{PV}}{dt} = \frac{1}{C_P}(I_{PV} - I_L) \\ \frac{dI_L}{dt} = \frac{1}{L}(-(1-\alpha)V_S + V_{PV}) \\ \frac{dV_S}{dt} = \frac{1}{C_S}((1-\alpha)I_L - I_{BAT}) \\ V_S = E_{BAT} + R_{BAT}I_{BAT} \end{cases} \tag{6}$$

where V_{PV}, V_S and I_L represent the average output voltage, the input voltage and the average inductor current during the switching period, respectively; L is the filter inductor; C_P is the input capacitor; C_S is the output capacitor and (E_{BAT}, R_{BAT}) is the load battery.

3. Controller Design and Stability Analysis

The block diagram of the proposed BS-SMC is shown in Figure 2. For each value of solar irradiation E and of cell temperature T, the block adaptive P&O-FLC enables the provision of an on-line calculation of the reference photovoltaic voltage V_{PV_REF}. The control signal α corresponds to the MPP and is generated from block BS-SMC controller.

3.1. Adaptive P&O–FLC

The differential power and the differential voltage are used as inputs of the adaptive P&O-FLC algorithm. ΔV_{PV_REF} is determined through an FL approach whose block diagram is presented in Figure 6 [33].

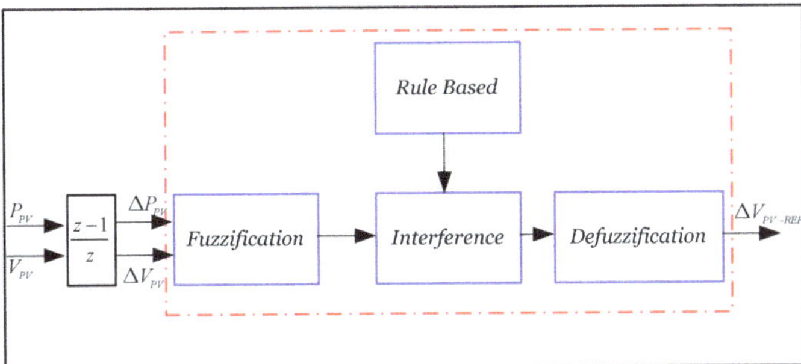

Figure 6. Block diagram of the adaptive perturb and observe fuzzy logic controller algorithm.

The functional diagram of the P&O-FLC is presented in Figure 7.

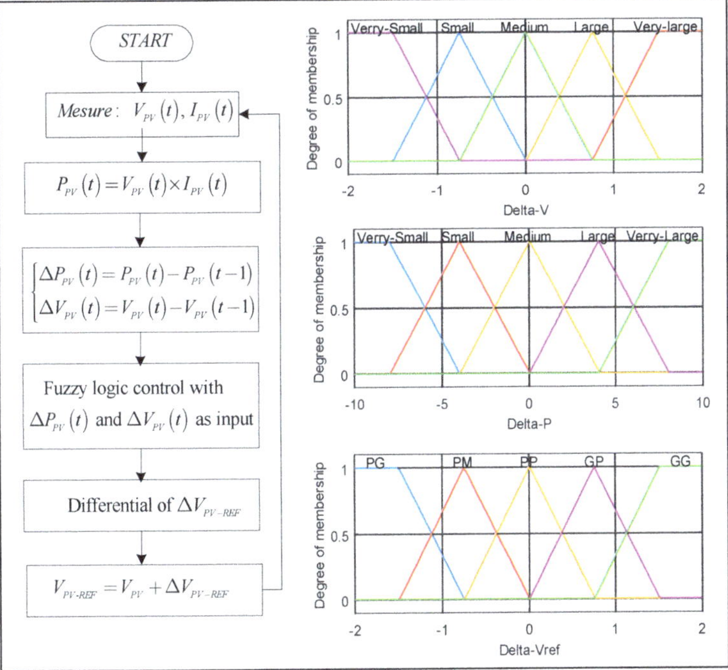

Figure 7. Flow chart of the adaptive P&O-FLC.

The five proposed variables are very small, small, medium, large and very large. The input membership function of the PV power ΔP_{PV} is the difference between the current power and the previous power. In addition, the difference between the current voltage and the previous voltage is the input membership function of the PV voltage ΔV_{PV}. Then, the difference between the current maximum PV output voltage and the previous maximum PV output voltage is the output membership function of the reference voltage ΔV_{PV_REF}. The two memberships functions ΔP_{PV} and ΔV_{PV} are converted to linguistic variables after their calculation. Then the output ΔV_{PV_REF} is generated by searching in a rule base table, consisting of 25 rules. A Mamdani's method is performed to determine the output of this algorithm [15,34,36,37].

3.2. Observer Design

The average model of the dc–dc boost converter can be deduced as follows:

$$\begin{cases} \dot{x}_1 = \frac{1}{L}(-(1-\alpha)V_S + x_2) \\ \dot{x}_2 = \frac{1}{C_P}(I_{PV} - x_1) \end{cases} \tag{7}$$

where x_1 represents the average inductor current and x_2 denotes the average PV voltage.

The state observer is designed based on the PV system module in order to estimate the state of the inductor current, which enables the application of the proposed BS-SMC MPPT. The estimated state vector $\hat{x} = [I_L, V_{PV}]^T$ is obtained by the state observer, in which the dynamics of the averaged state-space model should behave similarly to those of the real model.

The observer error between the real and the estimated states is defined as $\widetilde{x} = x - \hat{x}$ [21,22,28]. The proposed HGO can be written as:

$$\begin{cases} \dot{\hat{x}} = A\hat{x} + B\hat{x}u + M(y - \hat{y}) \\ \hat{y} = D\hat{x} \end{cases} \tag{8}$$

where $\dot{\hat{x}} = \left[\dot{\hat{I}}_L, \dot{\hat{V}}_{PV}\right]^T$, $\hat{y} = V_{PV}$ and $M = [m_1, m_2]^T$ denote an observer gain matrix chosen through analysing the stability of the proposed closed-loop system [38,39].

It can be noted that the HGO dynamics associated with the PV system are as follows:

$$\begin{cases} \dot{\hat{x}}_1 = -\frac{V_S}{L} + \frac{\hat{x}_2}{L} + \frac{V_S}{L}\alpha + m_1(y - \hat{y}) \\ \dot{\hat{x}}_2 = \frac{I_{PV}}{C_P} - \frac{\hat{x}_1}{C_P} + m_2(y - \hat{y}) \end{cases} \tag{9}$$

The voltage and current control errors between the reference and the estimated values are represented by defining the dynamic errors as $\widetilde{x} = x - \hat{x}$:

$$\begin{cases} \dot{\widetilde{x}}_1 = \frac{\widetilde{x}_2}{L} - m_1(y - \hat{y}) \\ \dot{\widetilde{x}}_2 = -\frac{\widetilde{x}_1}{L} - m_2(y - \hat{y}) \end{cases} \tag{10}$$

It can be observed from Equation (10) that the estimation error can be given in the following form:

$$\dot{\widetilde{x}} = (A - MD)\widetilde{x} \tag{11}$$

If $(A - MD)$ is a Hurwitz matrix, we can guarantee the convergence of asymptotic errors; consequently, $\underset{t \to \infty}{Lim}\widetilde{x}(t) = 0$.

The gains of the HGO can be selected as below:

$$\begin{cases} m_1 = \frac{k_1}{\varepsilon} \\ m_2 = \frac{k_2}{\varepsilon} \end{cases} \tag{12}$$

where $k_1 \gg k_2 \gg 1$ the constants k_1, k_2 and ε are definite positive.

3.3. Backstepping Sliding Mode Controller Design

The suggested MPPT controller is designed for the target that the voltage of the V_{PV} panel follows its reference value corresponding to the voltage at MPP of the PV panel [40]. This purpose is achieved by acting on the duty cycle α of the boost type dc–dc convertor.

Firstly, the voltage tracking error between V_{PV} and V_{PV_REF} is defined by:

$$e_1 = V_{PV} - V_{PV\_REF} \tag{13}$$

The derivative with regard to the time of e_1 is given by:

$$\dot{e}_1 = \dot{V}_{PV} - \dot{V}_{PV\_REF} = \frac{1}{C_P}(I_P - I_L) - \dot{V}_{PV\_REF} \tag{14}$$

Considering the first Lyapunov candidate function as follows:

$$V_1 = \frac{1}{2}e_1^2 \tag{15}$$

Time derivative of V_1 is given by:

$$\dot{V}_1 = e_1\dot{e}_1 = e_1\left(\frac{1}{C_P}(I_P - I_L) - \dot{V}_{PV\_REF}\right) \tag{16}$$

In the case that $\dot{V}_1$ is negative, we have:

$$\frac{1}{C_P}(I_P - I_L) - \dot{V}_{PV\_REF} = -Ke_1 \tag{17}$$

$$I_{L\_REF} = C_P\left(Ke_1 + \frac{I_P}{C_P} - \dot{V}_{PV\_REF}\right) \tag{18}$$

where K is a constant positive definite.

The second Lyapunov function is selected as follows:

$$V_2 = \frac{1}{2}e_1^2 + \frac{1}{2}s^2 \tag{19}$$

where S is the sliding surface given by:

$$s = \lambda_1 e_2 + \lambda_2 \int e_2 \tag{20}$$

$$e_2 = I_L - I_{L\_REF} \tag{21}$$

Consider the candidate Lyapunov function positive definite. The derivative with respect to time of the Lyapunov function is obtained through the following equation:

$$\begin{aligned}\dot{V}_2 &= \dot{e}_1 e_1 + s\dot{s} \\ &= -Ke_1^2 + s\left(\lambda_1\dot{e}_2 + \lambda_2 e_2\right)\end{aligned} \tag{22}$$

where $\dot{s}$ is given as follows:

$$\dot{s} = \lambda_1\dot{e}_2 + \lambda_2 e_2 = \frac{\lambda_1}{L}\left(-(1-\alpha)V_S + V_{PV}\right) - \lambda_1\dot{I}_{L\_REF} + \lambda_2 e_2 \tag{23}$$

We consider the dynamics of the sliding surface as follows [27]:

$$\dot{s} = q_0 s - q_1 sign(s) \tag{24}$$

where q_0 and q_1 are positive constants.

The expression for the control input deduced from Equations (23) and (24) by considering $s = \dot{s} = 0$, can be determined as:

$$\alpha = \frac{\frac{L}{\lambda_1}\left(-q_0 - q_1 sign(s) + \lambda_1 C_P\left(Ke_1 + \frac{I_P}{C_P} - \dot{V}_{PV\_REF}\right) - \lambda_2 e_2\right) - V_{PV}}{V_S} + 1 \tag{25}$$

3.4. Stability Analysis

To prove the stability of e_1 and e_2, we have resorted to the Barbalat's Lemma.

Barbalat's Lemma: *If the differentiable function $V(t)$ has a finite limit as $t \rightarrow \infty$ and if $\dot{V}(t)$ is uniformly continuous (or $\ddot{V}(t)$ is bounded), then we have $\dot{V}(t) \rightarrow 0$ as $t \rightarrow \infty$.*

By exploiting Equations (19), (22) and (24), the following equation can be obtained:

$$\begin{aligned}
\dot{V}_2 &= \dot{V}_1 + s\dot{s} = e_1\dot{e}_1 + s\dot{s} \\
&= -Ke_1^2 - s(q_0 s + q_1 sign(s)) \\
&\leq -Ke_1^2 - q_0 - q_1|s|
\end{aligned} \tag{26}$$

In order to make $\dot{V}_2 < 0$, the constants K, q_0 and q_1 should be positive.

4. Simulation Results

This section is dedicated to the simulation results of the BS-SMC scheme applied to the boost dc–dc converter for the PV storage system as exhibited in Figure 2. To demonstrate the superiority of the proposed BS-SMC, a comparison simulation with a conventional PI controller has been established under a MATLAB/Simulink package with the SimPower Toolbox. In order to point out the strengths and shortcomings of every controller, the two procedures are performed under similar tests conditions. The simulations were done with the sampling time of 100 μs for the global PV model. A monocrystalline Solo Line LX-100M model PV generator is used in this work with a peak of power of 100 W under standard test conditions (STC), such as a fixed value of solar radiation (E = 1 kW/m^2) and a fixed temperature (T = 25 °C). Table 1 shows the electrical parameters of the single PV, whereas the specifications of the controller parameters and the dc–dc boost converter are summarized in Table 2.

Table 2. Simulation parameters.

Parameter	Name	Value
C_P	Input capacitor	2200 μF
L	Inductor	5 mH
C_S	Output capacitor	4700 μF
f_s	MOSFET switching frequency	25 kHz
q_0	sliding surface	10
q_1	sliding surface	1
m_1	Gain of observer	60
m_2	Gain of observer	20

The storage device is a lead–acid battery of the PowerSafe TS range. It can reach 5200 cycles to a depth of discharge of 25%. The main characteristics of fully charged elements at a temperature of 25 °C are shown in Table 3.

Table 3. Lead acid battery parameters.

Parameter	Name	Value
C	Nominal capacity	390 Ah
R_{Bat}	Internal resistor	0.64*12(7.68 mΩ)
E_{Bat}	Nominal voltage	12*2 V(24 V)

The simulation results are evaluated according to three cases. The first case study is to evaluate the system performance of the proposed control during the system operation in STC. The second case consists of disrupting the climatic condition profile by comparing its performance with the PI controller technique. The third one is devoted to examining the function of the system in varying climatic conditions by introducing the observer studied.

Case 1: Standard operating conditions.

The first simulation section focused on the tracking performance under constant climatic conditions ($E = 1$ kW/m^2 and $T = 25\ °$C). The results of the voltage evolution and active power curves' response for the PV system are obtained by the proposed BS-SMC, as shown in Figure 8a, b. It can be seen from Figure 8b that the voltage follows its reference imposed by the adaptive P&O-FL MPPT technique with a fast setting time (around 27 ms), less dynamic error and any overshoot. It can be noted from Figure 8a that the MPPT based on BS-SMC applied in the boost dc–dc converter operates the power of the PV generator to MPP (100W). Furthermore, Figure 9a illustrates the performances of the inductor current with the presented control technique. Obviously, the proposed control using BS-SMC for the boost dc–dc converter stabilizes the PV output current to the optimal value with faster dynamic response, less overshoot and high precision and stability. Figure 10a,b show the convergence of the dynamic error and the sliding surface signal to zero. It can be concluded from this figure that the designed BS-SMC presents a good transition response, limits the chattering phenomenon and provides a good tracking performance.

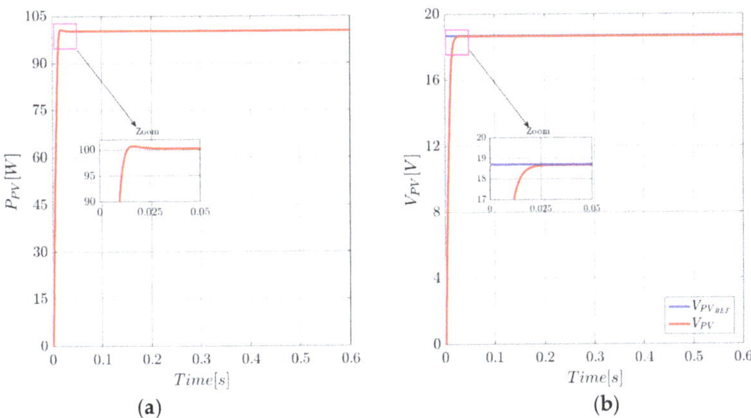

Figure 8. PV array output power and voltage under standard condition: (**a**) the PV power curve; (**b**) tracking of V_{PV} with respect to V_{PV_REF}.

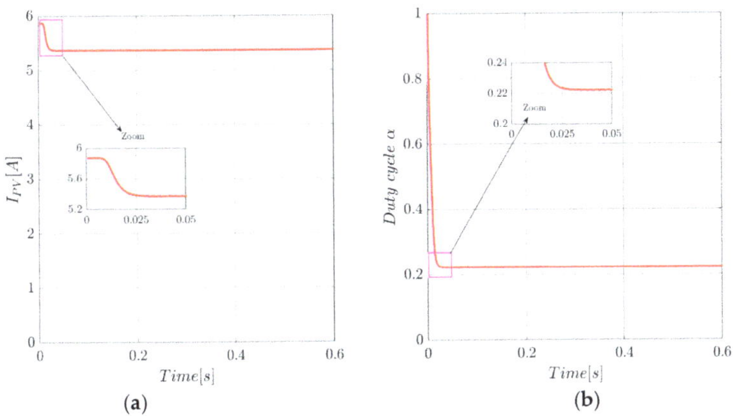

Figure 9. Maximum Power Point Tracking in standard operating conditions: (**a**) PV array output current; (**b**) duty cycle $\alpha(t)$.

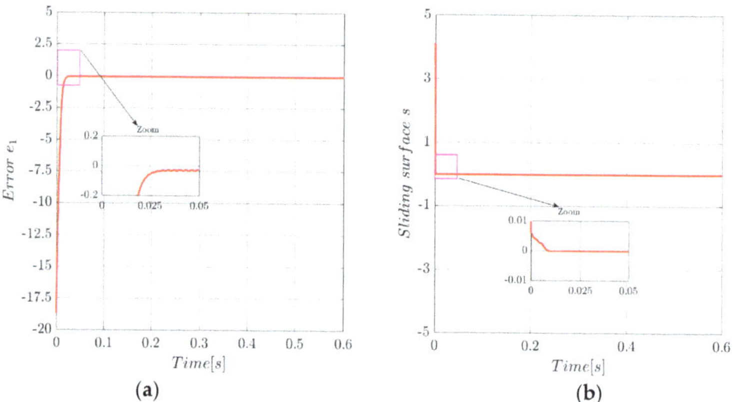

Figure 10. MPPT performance: (**a**) evolution of error e_1; (**b**) sliding surface s.

Case 2: Variation of solar irradiation and temperature.

The second simulation section is devoted to the control of the boost dc–dc converter using the proposed MPPT approach in case of changes in weather conditions, as shown in Figure 11a,b.

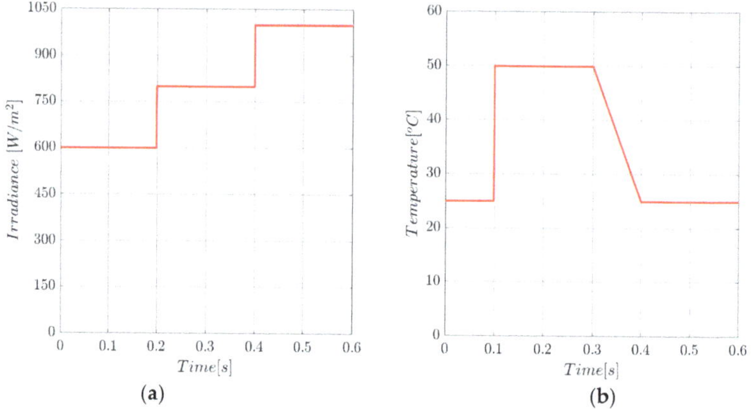

Figure 11. Varying level of weather conditions: (**a**) solar irradiance profile; (**b**) temperature profile.

The results of the evolution of voltage, current and power curves' responses are obtained by using a PI controller which has been compared with the proposed BS-SMC in Figures 12–14. It can be observed that the classical PI controller is dictated by the variation of solar irradiance and external disturbances, while the system controlled by the BS-SMC is more robust to variation in weather conditions. The proposed control ensures a better dynamic response and robustness under the solar irradiance changes during this test. The power, voltage and current results of the PV panel are presented in Figure 12a,b and Figure 13a, respectively.

According to Figure 12b, the P&O-FLC successfully generates the tracking of the MPP which is successfully tracked through the BS-SMC controller. Furthermore, in Figure 12a we can see that in specific situations when the irradiation rose from 0 to 600 W/m^2, the designed algorithm worked well with low solar irradiation and resulted in negligent power losses. It can also be observed that BS-SMC has a fast dynamic response with very low oscillation and stability.

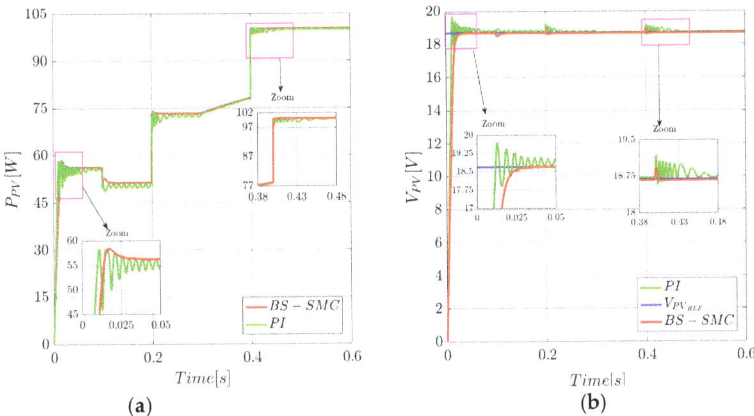

Figure 12. Comparison of tracking MPP of BS-SMC and PI controller under varying solar irradiance and temperature. (**a**) PV array output power; (**b**) tracking of V_{PV} with respect to V_{PV_REF}.

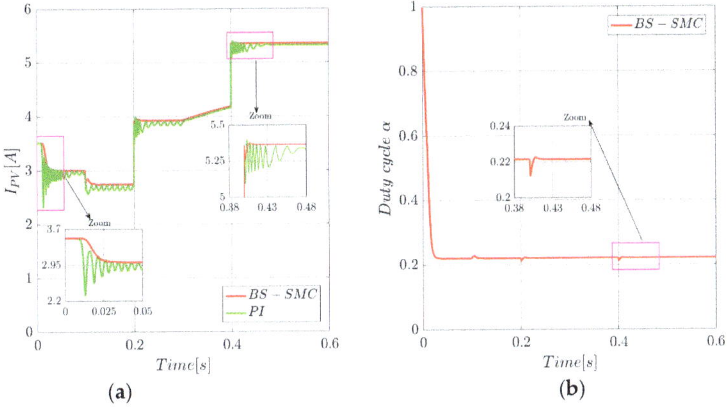

Figure 13. MPPT tracking under varying weather conditions: (**a**) PV array output current; (**b**) duty cycle $\alpha(t)$.

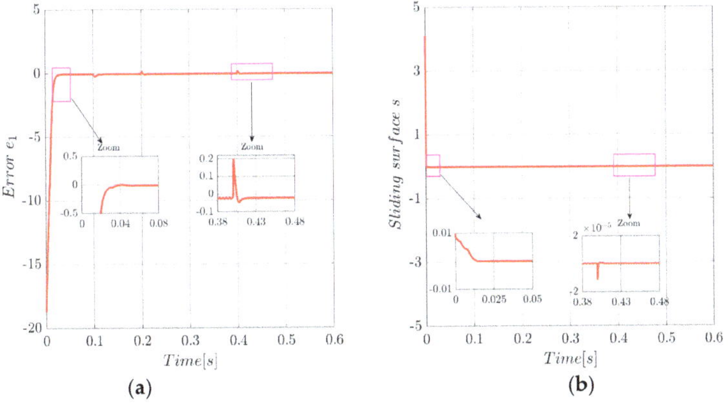

Figure 14. MPPT performance under a step change of the climatic conditions: (**a**) dynamic evolution of error e_1; (**b**) dynamic evolution of sliding surface s.

Figure 12a shows that the power increases under a different level of the solar irradiance and temperature, due to increases in the current and voltage of the panel. On the other hand, in order to confirm the performance of the BS-SMC approach, Figures 12–14 present a comparative study between the proposed controller and the conventional PI controller [9,17,40]. This comparison is according to weather condition as presented in Table 4.

Table 4. Performance comparison of the BS-SMC and PI controller.

Controller	Power Overshoot (%)	Power Ripple (W)	Response Time (ms)	Power Extraction Efficiency (%)
PI	1.68	6.1575	47.775	98
BS-SMC	1.6	2.8825	10.95	99.4

During this experiment, it is noteworthy that the system follows the reference under 10.95 ms faster with a steady state error to zero using the BS-SMC where there are strong oscillations in the PI controller. In addition, the point at maximum power is reached with almost negligible ripple in less than 2.8825 W. Moreover, the boost dc–dc converter successfully extracts maximum power with more than 99.4% efficiency. Also, when the solar irradiance and temperature are kept at 1 kW/m^2 and 25 °C, successively, the average output power is equal to 100.1 W. Compared with the PI controller, the power increases by about 2% of the value obtained by the proposed controller. Moreover, these comparisons confirm the relevance and benefit of the presented control strategies in terms of the voltage monitoring of the PV model at the MPP.

A zoomed view of the oscillations around the MPP using the PI controller is presented in Figure 12b. It is clear from this figure that the ripples output voltage of the proposed controller is much lower than that of the PI controller.

According to climatic condition variation, the proposed BS-SMC controller proves its robustness, which reduces the chatter phenomenon, as shown in Figure 14b. In fact, the strategy proposed by BS-SMC offers better dynamic performance than that used by the conventional PI control method.

Case 3: Application of HGO under solar irradiation and temperature variation.

In this case, the performance of the presented BS-SMC at tracking the MPP without the use of inductor current sensors is evaluated as a function of solar irradiance and temperature variations.

The profiles of the climatic conditions in this simulations study are presented in Figure 11a,b.

Figures 15–17 show the dynamic performance of the tracker at varying irradiance and temperature. It is clearly shown from Figure 15a that the designed sensor-less MPPT does not exhibit much oscillation around the MPP.

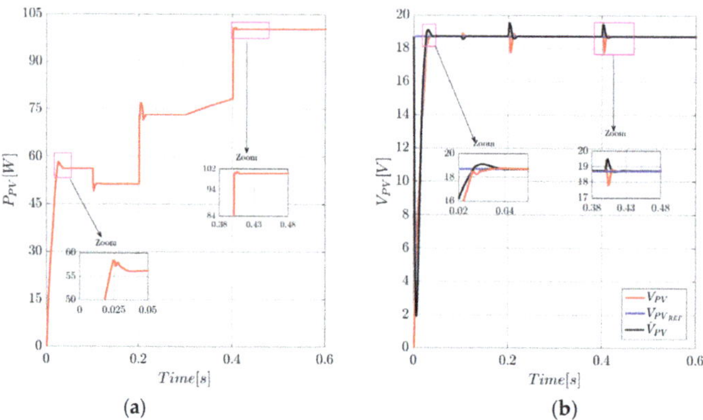

Figure 15. PV array output power and voltage under varying solar irradiance and temperature without sensor inductor current: (**a**) PV array output power; (**b**) tracking of V_{PV} with respect to V_{PV_REF}.

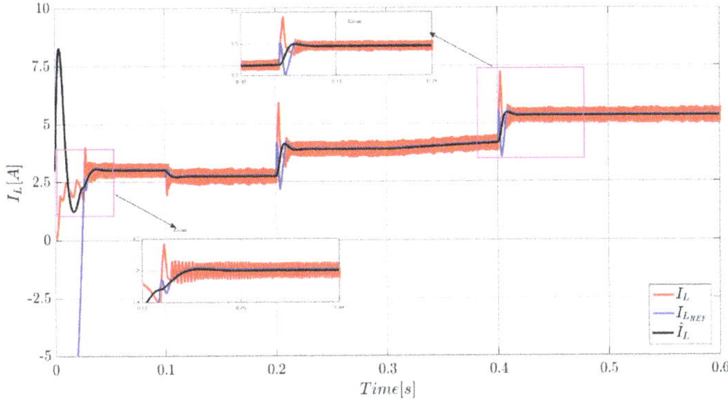

Figure 16. Inductor current without sensors.

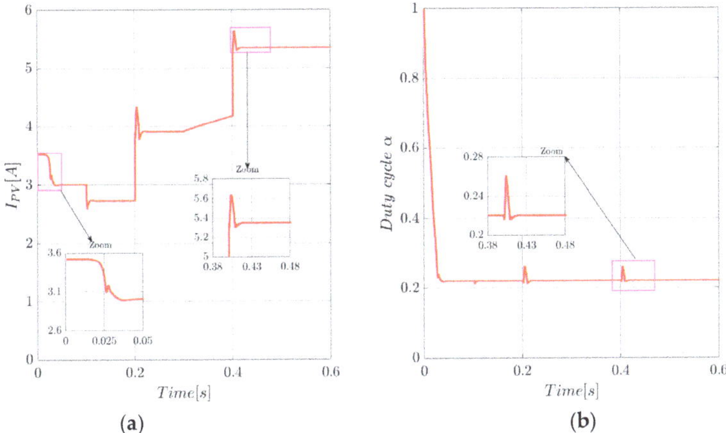

Figure 17. MPPT tracking under conditions climatic variations without sensor inductor current: (**a**) PV array output current; (**b**) duty cycle $\alpha(t)$.

For more details, the key figures obtained for the BS-SMC controller without inductor current sensors are shown in Table 5.

Table 5. Performance of the BS-SMC without sensor inductor current.

Algorithm	Power Overshoot (%)	Power Ripple (W)	Response Time (ms)	Power Extraction Efficiency (%)
BS-SMC	3.44	2.745	9.75	99.79

Figure 16 shows the error between the inductor current and it is estimated to be almost equal zero.

Figure 15b shows that when the estimation state starts in the observer reel state, there is an instantaneous estimation error of the voltage which is rapidly reduced to zero during the estimation process.

In Figure 16, it can be noted that the HGO has better performance by comparing the measurements of the inductor current with the estimation. The choice of gains m_1 and m_2 confirm the correct operation of the observers as the estimated current correctly tracks the real state in less than 0.036 s, under an initial solar irradiance and temperature of 600 W/m^2 and 25 °C, respectively.

In Figure 15b, we can see that the transient response time is very short when a sudden change in solar irradiance and temperature occur at t = 9.75 ms, revealing the best dynamic performance of the

BS-SMC approach. It can be noticed that, at the moment of the step change of the weather conditions, the voltage of the PV generator can be stabilized to its reference value. In addition, the BS-SMC strategy provides a stable power proportional to the irradiation and temperature levels, at a response time of the order of 1 ms when the solar irradiation E rises from 800 W/m^2 to 1000 W/m^2 with the temperature maintained at 25 °C. Figure 15a presents the power of the PV module. From this figure, some fluctuations are recorded which are ascribed to the dynamics of the observer.

As seen in Figure 15a,b, the MPP is successfully tracked and the PV voltage is stabilized at the reference value. Hence, it is obvious, by examining Figures 15b and 16, that both the PV voltage and the inductor current have smaller overshoots. For example, when the solar irradiance and temperature are kept at 1 kW/m^2 and 25 °C, respectively, in this case, $m_1 = 60$ and $m_2 = 20$ the estimated inductor current and the PV voltage have an overshoot of approximately 0.03 A and 0.9 V, respectively. It can be seen that the proposed state observer-based control algorithm is able to force the PV generator to operate at MPP using only three sensors to measure voltage and current.

In the sliding surface mode, a phenomenon known as chattering can take place. It is manifested by high-frequency switching around the sliding surface. These commutations can excite unwanted dynamics that may destabilize damage or even destroy the studied system.

This chattering phenomenon is limited by the introduced BS-SMC controller, as shown in Figure 18 which demonstrates the efficiency of this approach.

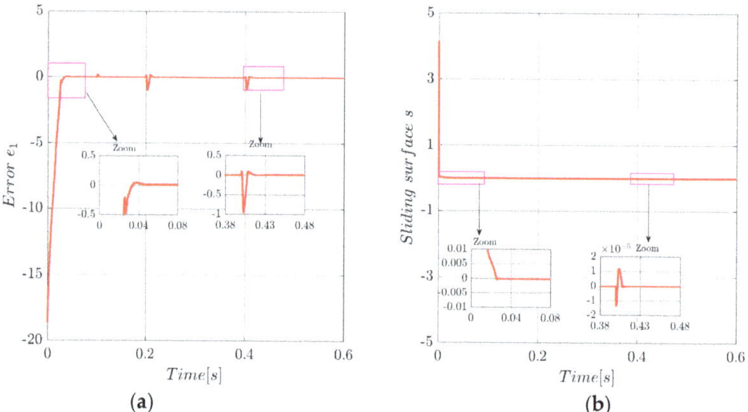

Figure 18. MPPT performance under a step change in the solar irradiance and temperature without sensor inductor current: (**a**) dynamic evolution of error e_1; (**b**) dynamic evolution of sliding surface s.

5. Experimental Results and Discussion

An experimental model has been presented and analyzed on a more sophisticated test bench for a standalone PV system in the CRTEn research laboratory located in the technology park of Borj Cedrya Tunisia at latitude 36.717° and longitude 10.427°. The prototype built consists of a BP Solar LX-100M photovoltaic panel, a dc–dc boost converter and a lead acid type battery load, as illustrated in Figure 19. The PV storage system is controlled by a Control Engineering DS 1104 board through a Matlab/Simulink environment.

The block diagram of the proposed BS-SMC approach has been implemented to generate the pulse with modulation (PWM) signal for acting on the MOSFET gate of the dc–dc boost converter, as seen in Figure 20. The parameters of the PV storage system are presented in Table 2 and the switching frequency of the boost converter is chosen to be 25 kHz. The sampling time of the system is chosen with the performance of controller board at $T_S = 10^{-4}$ s. The BP Solar LX-100M PV panel with an angle of 37° has the parameters presented in Table 1.

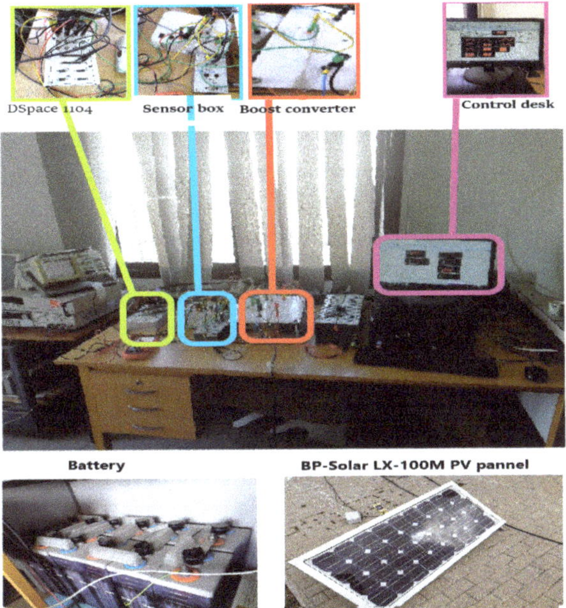

Figure 19. Experimental test bench.

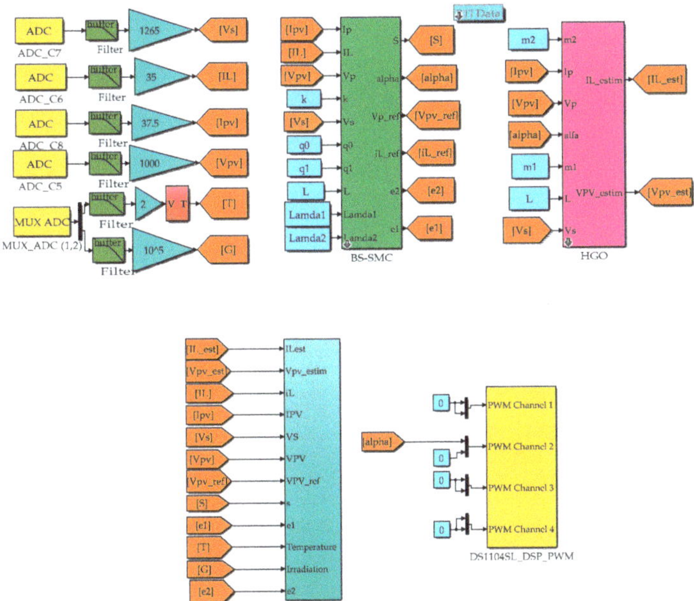

Figure 20. Block diagram implementation of the proposed BS-SMC MPPT.

The data acquisition of the input voltage and the input current for the test bench comes from LEM sensors LV25-P and LA25-P, respectively. These serve as inputs to the interface of the dSPACE 1104 while the PWM output signal generated from the DS 1104 varies between 0 and 5 V. To amplify the output voltage to a voltage sufficient to the MOSFET (10V), an amplifier circuit is constructed. When the pulses are amplified, they go through an isolated circuit to separate the supply circuit and the control circuit. Control Desk software is used to monitor the displacement of the MPP on the P–V characteristics under changing climatic conditions. To reduce unwanted high-frequency noise from current and voltage measurements during acquisition, low-pass filters have been designed. Given the absence of a PV emulator, the results are presented based on real-time temperature and solar irradiation measurement data, as shown in Figures 21 and 22.

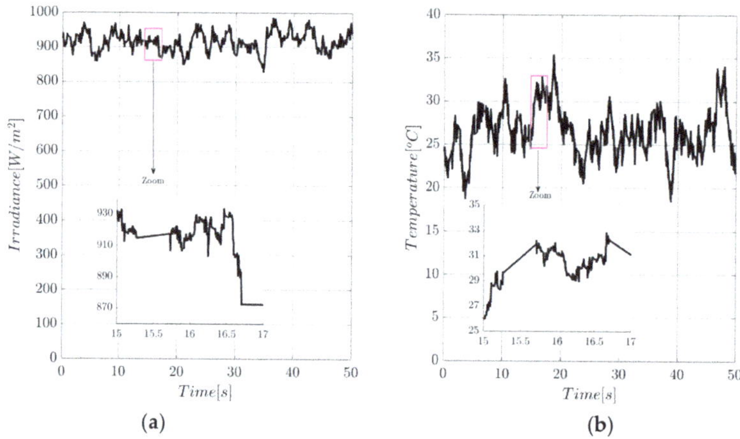

Figure 21. Experimental results of climatic conditions variation: (**a**) experimental solar irradiance profile; (**b**) experimental temperature profile.

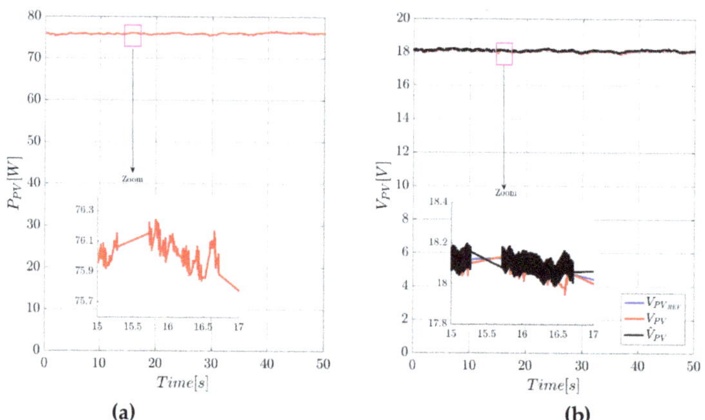

Figure 22. Experimental PV array output power and voltage under varying solar irradiance and temperature without sensor inductor current: (**a**) PV array output power; (**b**) tracking of V_{PV} with respect to V_{PV_REF}.

The block diagram implantation of the proposed approach given in Figure 2 is used to confirm the real time simulation. It can be seen that the proposed MPPT has been implemented using the DS1104 platform.

To evaluate and verify the efficiency of the proposed MPPT method based on the BS-SMC controller, experimental results were obtained in the Dspace 1104 card. Figures 22–24 show the experimental results of the voltage, current and power of the PV module and inductor current by using the proposed MPPT approach.

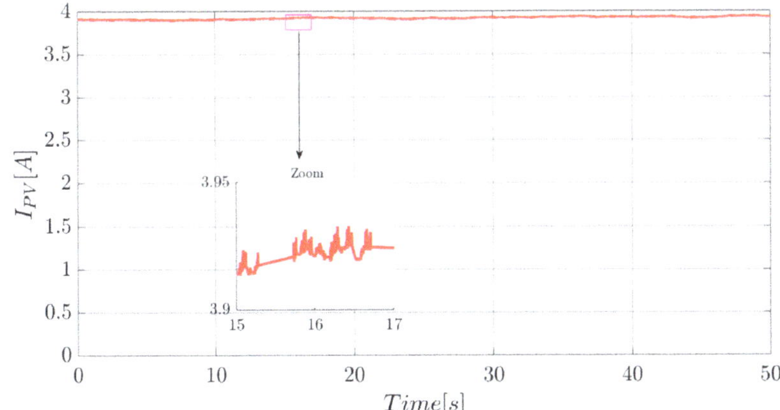

Figure 23. Experimental results of the output PV current.

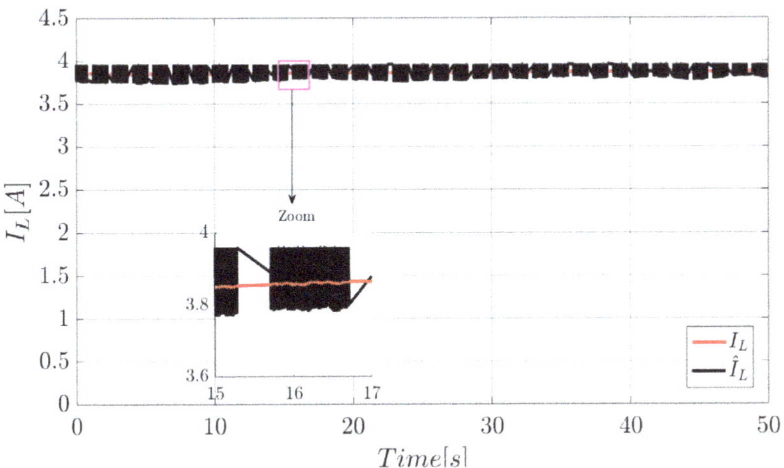

Figure 24. Experimental results of the inductor current estimation.

During the interval time $t = [15, 17]$ s, the maximum irradiance and temperature level are $E = 934.1 \text{W/m}^2$ and T = 30.18 °C, respectively. In this condition, the optimal current and voltage of the PV module are about 3.92 A and 18.24 V, respectively. Then, the maximum PV power is around 77.51 W. Table 6 shows clearly that the presented scheme is characterized by negligible overshoot and permanent low voltage ripples comparing with the results obtained in [36].

Table 6. Experiment performance of the BS-SMC.

	Power Overshoot	Power Ripple
PV power	1.5%	2.2 W
PV voltage	0.7%	0.32 V

It can be noted from these results that the MPP is always achieved. The validation of the robustness of the proposed BS-SMC in the presence of temperature and solar irradiation variation is guaranteed. From the experimental measurements presented in Figures 22 and 23, we can see small oscillations around the average value of the PV output voltage, power and current, respectively. The characteristics P_V and I_V curves of the PV module under weather conditions' variation are shown in Figure 25. This confirms that the MPP voltage is only affected by fast-varying solar irradiance.

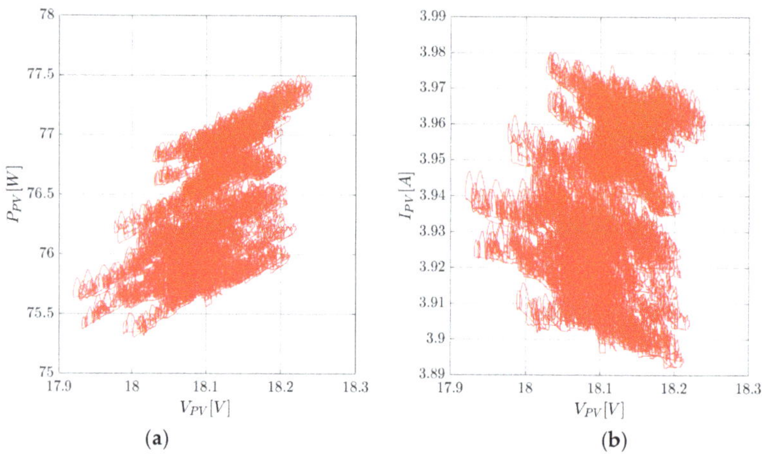

Figure 25. PV array output characteristic curves: (**a**) $P_{PV} - V_{PV}$ characteristic curves of the PV module; (**b**) $I_{PV} - V_{PV}$ characteristic curves of the PV module.

According to the experimental results, it is concluded that, under varying irradiation and temperature conditions, the system using the proposed approach is able to operate MPP accurately, and offers better stability and insignificant steady state error. Interestingly, the observed inductor current has some minimal error between the experimental results and the estimated state by the HGO, as illustrated in Figure 24.

6. Conclusions

In this paper, we have presented and analyzed a robust BS-SMC controller to track the MPP of a stand-alone PV system. The proposed approach is performed to regulate the PV module output voltage to its reference trajectory obtained by the adaptive P&O-FLC technique. The maximum PV output power is achieved under changing climatic conditions. The Lyapunov stability analysis is used to verify the global asymptotic stability of the PV system. The simulation and experimental results under changing climatic conditions show the performance and the robustness of the MPPT controller. In addition, the presented technique performs better when compared with the classical PI controller. In fact, the simulation and experimental results confirm the efficiency of the HGO. The BS-SMC controller is simple to design and easy to implement in a real-time application. The major contributions of our study are as follows: reducing the required sensor by using HGO, reducing the chattering phenomenon by using the BS-SMC approach, the successful implementation of the BS-SMC controller in a dSPACE1104 card and the successful application of the dSPACE-based controller to control the PV storage application with a robust control performance. Future work can focus on the application of the proposed BS-SMC to other dc–dc converters with battery loads such as buck converters; and discretization of the proposed MPPT controller can also be investigated. Work is currently in progress and the results will be reported in due course.

Author Contributions: M.B. proposed the main idea, performed the investigation and developed the software; M.M., B.K. and R.B. collaborated to achieve the work; A.S. led the project and acquired the funds for research.

Funding: This research was funded by LANSER Laboratory/CRTEn B.P.95 Hammam-Lif 2050, Tunis, Tunisia.

Conflicts of Interest: The authors declare no conflict of interest.

References

1. Zaouche, F.; Rekioua, D.; Gaubert, J.-P.; Mokrani, Z. Supervision and control strategy for photovoltaic generators with battery storage. *Int. J. Hydrogen Energy* **2017**, *42*, 19536–19555. [CrossRef]
2. Mohapatra, A.; Nayak, B.; Das, P.; Mohanty, K.B. A review on MPPT techniques of PV system under partial shading condition. *Renew. Sustain. Energy Rev.* **2017**, *80*, 854–867. [CrossRef]
3. Ali, N.; Armghan, H.; Ahmad, I.; Armghan, A.; Khan, S.; Arsalan, M. Backstepping based non-linear control for maximum power point tracking in photovoltaic system. *Sol. Energy* **2018**, *159*, 134–141.
4. Azzollini, I.A.; Di Felice, V.; Fraboni, F.; Cavallucci, L.; Breschi, M.; Rosa, A.D.; Zini, G. Lead-Acid Battery Modeling Over Full State of Charge and Discharge Range. *IEEE Trans. Power Syst.* **2018**, *33*, 6422–6429. [CrossRef]
5. Spanos, C.; Turney, D.E.; Fthenakis, V. Life-cycle analysis of flow-assisted nickel zinc-, manganese dioxide-, and valve-regulated lead-acid batteries designed for demand-charge reduction. *Renew. Sustain. Energy Rev.* **2015**, *43*, 478–494. [CrossRef]
6. Rezkallah, M.; Hamadi, A.; Chandra, A.; Singh, B.; Hamadi, A. Design and Implementation of Active Power Control With Improved P&O Method for Wind-PV-Battery-Based Standalone Generation System. *IEEE Trans. Ind. Electron.* **2018**, *65*, 5590–5600.
7. Kihal, A.; Krim, F.; Laib, A.; Talbi, B.; Afghoula, H. An improved MPPT scheme employing adaptive integral derivative sliding mode control for photovoltaic systems under fast irradiation changes. *ISA Trans.* **2018**, *87*, 297–306. [CrossRef]
8. Zakzouk, N.E.; Williams, B.W.; Helal, A.A.; Elsaharty, M.A.; Abdelsalam, A.K. Improved performance low-cost incremental conductance PV MPPT technique. *IET Renew. Power Gener.* **2016**, *10*, 561–574. [CrossRef]
9. Bounechba, H.; Bouzid, A.; Snani, H.; Lashab, A. Real time simulation of MPPT algorithms for PV energy system. *Int. J. Electr. Power Energy Syst.* **2016**, *83*, 67–78. [CrossRef]
10. Koohi-Kamali, S.; Rahim, N.; Mokhlis, H. Smart power management algorithm in microgrid consisting of photovoltaic, diesel, and battery storage plants considering variations in sunlight, temperature, and load. *Energy Convers. Manag.* **2014**, *84*, 562–582. [CrossRef]
11. Messalti, S.; Harrag, A.; Loukriz, A. A new variable step size neural networks MPPT controller: Review, simulation and hardware implementation. *Renew. Sustain. Energy Rev.* **2017**, *68*, 221–233. [CrossRef]
12. Mhamed, F.; Mohamed Larbi, E.; Smail, Z. Hardware implementation of the fuzzy logic MPPT in an Arduino card using a Simulink Support Package for photovoltaic application. *IET Renew. Power Gener.* **2018**, *13*, 510–518.
13. Reddy, K.J.; Sudhakar, N. ANFIS-MPPT control algorithm for a PEMFC system used in electric vehicle applications. *Int. J. Hydrogen Energy* **2019**, *44*, 15355–15369. [CrossRef]
14. Shah, N.; Rajagopalan, C. Experimental evaluation of a partially shaded photovoltaic system with a fuzzy logic-based peak power tracking control strategy. *IET Renew. Power Gener.* **2016**, *10*, 98–107. [CrossRef]
15. Farajdadian, S.; Hosseini, S.H. Design of an optimal fuzzy controller to obtain maximum power in solar power generation system. *Sol. Energy* **2019**, *182*, 161–178. [CrossRef]
16. Ahmad Khateb, M.N. Fuzzy logic controller based sepic converter for maximum power point tracking. *IEEE Trans. Ind. Appl.* **2016**, *50*, 2349–2358. [CrossRef]
17. El Khazane, J.; Tissir, E.H. Achievement of MPPT by finite time convergence sliding mode control for photovoltaic pumping system. *Sol. Energy* **2018**, *166*, 13–20. [CrossRef]
18. Ya-Ting, L.; Chian-Song, C.; Tse-Wei, C. Maximum power point tracking of grid-tied photovoltaic power systems. In Proceedings of the 2014 International Power Electronics Conference, Hiroshima, Japan, 18–21 May 2014; pp. 440–444.
19. Yatimi, H.; Aroudam, E. Assessment and control of a photovoltaic energy storage system based on the robust sliding mode MPPT controller. *Sol. Energy* **2016**, *139*, 557–568. [CrossRef]

20. Yasin, A.R.; Ashraf, M.; Bhatti, A.I. A Novel Filter Extracted Equivalent Control Based Fixed Frequency Sliding Mode Approach for Power Electronic Converters. *Energies* **2019**, *12*, 853. [CrossRef]
21. Yang, B.; Yu, T.; Shu, H.; Zhu, D.; An, N.; Sang, Y.; Jiang, L. Perturbation observer based fractional-order sliding-mode controller for MPPT of grid-connected PV inverters: Design and real-time implementation. *Control Eng. Pract.* **2018**, *79*, 105–125. [CrossRef]
22. Keshari, K.; Neeli, S.; Vijayakumar, K. Design of a sliding-mode-controlled dc-dc converter for MPPT in grid-connected PV System. In Proceedings of the 2017 IEEE International Conference on Electrical, Instrumentation and Communication Engineering (ICEICE), Karur, India, 27–28 April 2017; pp. 1–8.
23. Pradhan, R.; Subudhi, B. Double Integral Sliding Mode MPPT Control of a Photovoltaic System. *IEEE Trans. Control Syst. Technol.* **2016**, *24*, 1. [CrossRef]
24. Belkaid, A.; Gaubert, J.-P.; Gherbi, A. Design and implementation of a high-performance technique for tracking PV peak power. *IET Renew. Power Gener.* **2017**, *11*, 92–99. [CrossRef]
25. Cortajarena, J.A.; Barambones, O.; Alkorta, P.; De Marcos, J. Sliding mode control of grid-tied single-phase inverter in a photovoltaic MPPT application. *Sol. Energy* **2017**, *155*, 793–804. [CrossRef]
26. Montoya, D.G.; Ramos-Paja, C.A.; Giral, R.; Ramos, C. Improved design of sliding mode controllers based on the requirements of MPPT techniques. *IEEE Trans. Power Electron.* **2016**, *31*, 1. [CrossRef]
27. Dahech, K.; Allouche, M.; Damak, T.; Tadeo, F. Backstepping sliding mode control for maximum power point tracking of a photovoltaic system. *Electr. Power Syst. Res.* **2017**, *143*, 182–188. [CrossRef]
28. Das, D.; Madichetty, S.; Singh, B.; Mishra, S. Luenberger Observer Based Current Estimated Boost Converter for PV Maximum Power Extraction-A Current Sensorless Approach. *IEEE J. Photovolt.* **2019**, *9*, 278–286. [CrossRef]
29. Kumari, S. Decentralized SoC based Droop with Sliding Mode Current Controller for DC Microgrid. In Proceedings of the 2018 5th IEEE Uttar Pradesh Section International Conference on Electrical, Electronics and Computer Engineering (UPCON), Gorakhpur, India, 2–4 November 2018.
30. Chen, F.; Jiang, R.; Zhang, K.; Jiang, B.; Tao, G. Robust Backstepping Sliding Mode Control and Observer-based Fault Estimation for a Quadrotor UAV. *IEEE Trans. Ind. Electron.* **2016**, *63*, 5044–5056. [CrossRef]
31. Arsalan, M.; Iftikhar, R.; Ahmad, I.; Hasan, A.; Sabahat, K.; Javeria, A. MPPT for photovoltaic system using nonlinear backstepping controller with integral action. *Sol. Energy* **2018**, *170*, 192–200. [CrossRef]
32. Kchaou, A.; Naamane, A.; Koubaa, Y.; M'Sirdi, N. Second order sliding mode-based MPPT control for photovoltaic applications. *Sol. Energy* **2017**, *155*, 758–769. [CrossRef]
33. Bjaoui, M.; Khiari, B.; Benadli, R.; Memni, M.; Sellami, A. Maximum Power Point Tracking of a photovoltaic pumping system with PSO controller design. In Proceedings of the 4th International Conference on Green Energy and Environmental Engineering (GEEE), Sousse, Tunisia, 22–24 April 2017.
34. Abu Eldahab, Y.E.; Saad, N.H.; Zekry, A. Enhancing the design of battery charging controllers for photovoltaic systems. *Renew. Sustain. Energy Rev.* **2016**, *58*, 646–655. [CrossRef]
35. Viloria-Porto, J.; Robles-Algarín, C.; Restrepo-Leal, D. A Novel Approach for an MPPT Controller Based on the ADALINE Network Trained with the RTRL Algorithm. *Energies* **2018**, *11*, 3407. [CrossRef]
36. Zainuri, M.A.A.M.; Rahim, N.A.; Soh, A.C.; Radzi, M.A.M. Development of adaptive perturb and observe-fuzzy control maximum power point tracking for photovoltaic boost dc–dc converter. *IET Renew. Power Gener.* **2014**, *8*, 183–194. [CrossRef]
37. Yilmaz, U.; Kircay, A.; Borek+ci, S. PV system fuzzy logic MPPT method and PI control as a charge controller. *Renew. Sustain. Energy Rev.* **2018**, *81*, 994–1001. [CrossRef]
38. Cho, H.; Yoo, S.J.; Kwak, S. State observer based sensor less control using Lyapunov's method for boost converters. *IET Power Electron.* **2015**, *8*, 11–19. [CrossRef]
39. Cimini, G.; Ippoliti, G.; Orlando, G.; Longhi, S.; Miceli, R. A unified observer for robust sensorless control of DC–DC converters. *Control Eng. Pract.* **2017**, *61*, 21–27. [CrossRef]
40. Anto, E.K.; Asumadu, J.A.; Okyere, P.Y. PID control for improving P&O-MPPT performance of a grid-connected solar PV system with Ziegler-Nichols tuning method. In Proceedings of the 2016 IEEE 11th Conference on Industrial Electronics and Applications (ICIEA), Hefei, China, 5–7 June 2016; pp. 1847–1852.

Article

Advanced MPPT Algorithm for Distributed Photovoltaic Systems

Hyeon-Seok Lee [1] and Jae-Jung Yun [2,*]

[1] Department of Electrical Engineering, POSTECH, Pohang 37673, Korea; hsasdf@postech.ac.kr
[2] Department of Electronics and Electrical Engineering, Daegu University, Gyeongsan 38453, Korea
* Correspondence: jjyun@daegu.ac.kr; Tel.: +82-53-850-6612

Received: 22 August 2019; Accepted: 18 September 2019; Published: 19 September 2019

Abstract: The basic and adaptive maximum power point tracking algorithms have been studied for distributed photovoltaic systems to maximize the energy production of a photovoltaic (PV) module. However, the basic maximum power point tracking algorithms using a fixed step size, such as perturb and observe and incremental conductance, suffer from a trade-off between tracking accuracy and tracking speed. Although the adaptive maximum power point tracking algorithms using a variable step size improve the maximum power point tracking efficiency and dynamic response of the basic algorithms, these algorithms still have the oscillations at the maximum power point, because the variable step size is sensitive to external factors. Therefore, this paper proposes an enhanced maximum power point tracking algorithm that can have fast dynamic response, low oscillations, and high maximum power point tracking efficiency. To achieve these advantages, the proposed maximum power point tracking algorithm uses two methods that can apply the optimal step size to each operating range. In the operating range near the maximum power point, a small fixed step size is used to minimize the oscillations at the maximum power point. In contrast, in the operating range far from the maximum power point, a variable step size proportional to the slope of the power-voltage curve of PV module is used to achieve fast tracking speed under dynamic weather conditions. As a result, the proposed algorithm can achieve higher maximum power point tracking efficiency, faster dynamic response, and lower oscillations than the basic and adaptive algorithms. The theoretical analysis and performance of the proposed algorithm were verified by experimental results. In addition, the comparative experimental results of the proposed algorithm with the other maximum power point tracking algorithms show the superiority of the proposed algorithm.

Keywords: PV system; P&O; INC; adaptive; DC–DC converter; DMPPT

---

1. Introduction

The renewable energy resources including wind power, biomass, solar heating, solar photovoltaic (PV), hydroelectric energy, and fuel cells have been widely used to reduce global warming effects caused by greenhouse gas emission [1–4]. Among these energy resources, solar PV has attracted attention as a promising renewable energy source due to the following reasons:

1. Diverse applications: PV system can be easily applied to microgrid, households, and buildings [5–7].
2. Low maintenance costs: Only PV module, inverter, and cable are required for maintenance and the maintenance period is long [8–10].
3. Technology development: Various methods were constantly introduced to improve the technology in the solar power industry [8–10].

PV systems consist of PV modules for converting sunlight into direct current (DC) electricity, as well as PV inverters for converting DC into alternating current (AC). Based on the connection method between the PV module and the PV inverter, the PV inverter can be categorized as a central inverter or as a module-level power electronic (MLPE). In the past, central inverters (Figure 1a) connected with series-connected PV modules were widely used because they have the advantages of simple structure and low cost [11]. However, central inverters suffered from significant performance degradation under partial shading due to multiple maximum power points and mismatches in PV modules. To solve the partial shading problem, MLPEs (Figure 1b), which include a DC-optimizer and a micro-inverter, were introduced [12–16]. MLPEs are connected to one PV module and harvest optimum power by performing module-level maximum power point tracking (MPPT); this is also known as distributed MPPT (DMPPT) [17–19]. Thereby, MLPEs solve the partial shading problem and improve the performance of an entire PV system.

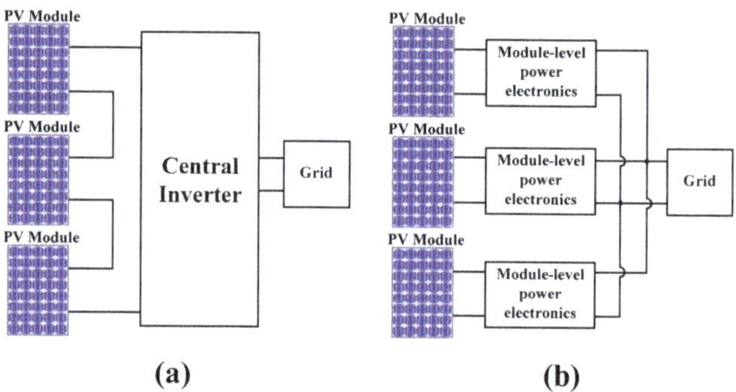

Figure 1. Block diagrams of (**a**) the central inverter and (**b**) the module-level power electronics.

Among many MPPT algorithms, hill-climbing, perturb and observe (P&O), and incremental conductance (INC) have been widely used due to their simplicity and ease of implementation [20–24].

Hill-climbing and P&O algorithms operate on the same fundamental principle that the variation (ΔV_{PV}) of PV voltage (V_{PV}) and the variation (ΔP_{PV}) of PV power (P_{PV}) become zero at the maximum power point (MPP). The difference between the P&O and hill-climbing algorithms is that P&O uses a PI controller. Operations in both hill-climbing and P&O algorithms can be classified into five modes and are described in Figure 2 and Table 1 [20–23]. The INC algorithm tracks the MPP based on the principle that the variation (ΔI_{PV}) of PV current (I_{PV}) becomes zero and the slope of $-\Delta I_{PV}/\Delta V_{PV}$ is the same as I_{PV}/V_{PV} at the MPP. The INC algorithm also has five operating modes, and its principle of operation is described in Figure 3 and Table 2 [24]. However, these basic MPPT algorithms have a trade-off problem between tracking accuracy and speed because fixed step size is used for perturbation. If small fixed step size is used, tracking accuracy increases, but speed is slower. In this case, the basic MPPT algorithms can fail to track the MPP under dynamic weather conditions. At large fixed step size, tracking speed is faster, but tracking accuracy decreases, which causes a low MPPT efficiency.

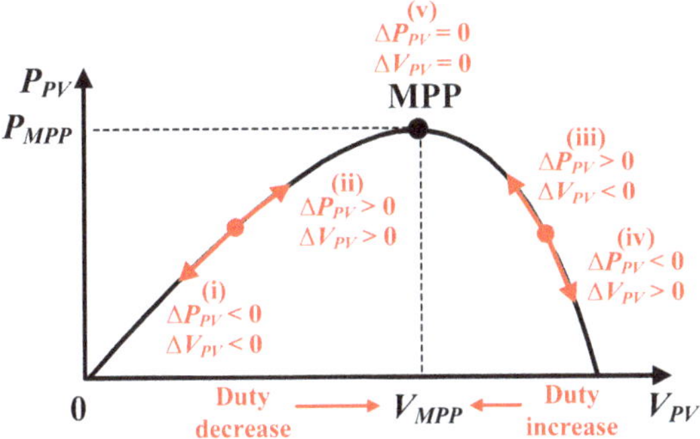

Figure 2. Principle of operation for the hill-climbing and perturb and observe (P&O) algorithms.

Table 1. Methodology of the hill-climbing and P&O algorithms.

Conditions	Actions
(i) $\Delta P_{PV} < 0$ and $\Delta V_{PV} < 0$	Duty decrease
(ii) $\Delta P_{PV} > 0$ and $\Delta V_{PV} > 0$	Duty decrease
(iii) $\Delta P_{PV} > 0$ and $\Delta V_{PV} < 0$	Duty increase
(iv) $\Delta P_{PV} < 0$ and $\Delta V_{PV} > 0$	Duty increase
(v) $\Delta P_{PV} = 0$ and $\Delta V_{PV} = 0$	No action

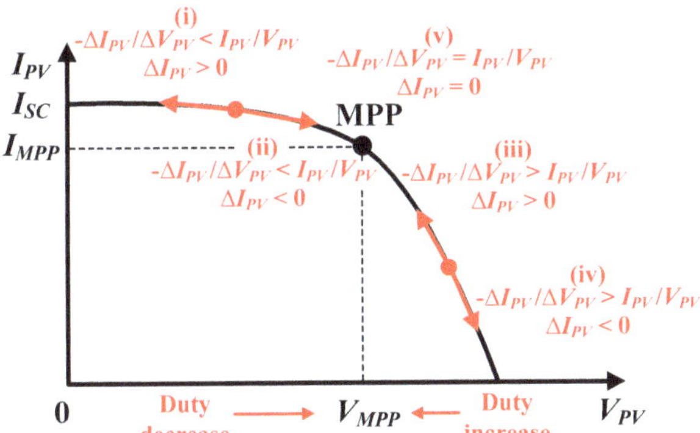

Figure 3. Principle of operation for the incremental conductance (INC) algorithm.

Table 2. Methodology of the INC algorithm.

Conditions	Actions
(i) $-\Delta I_{PV}/\Delta V_{PV} < I_{PV}/V_{PV}$ and $\Delta I_{PV} > 0$	Duty decrease
(ii) $-\Delta I_{PV}/\Delta V_{PV} < I_{PV}/V_{PV}$ and $\Delta I_{PV} < 0$	Duty decrease
(iii) $-\Delta I_{PV}/\Delta V_{PV} > I_{PV}/V_{PV}$ and $\Delta I_{PV} > 0$	Duty increase
(iv) $-\Delta I_{PV}/\Delta V_{PV} > I_{PV}/V_{PV}$ and $\Delta I_{PV} < 0$	Duty increase
(v) $-\Delta I_{PV}/\Delta V_{PV} = I_{PV}/V_{PV}$ and $\Delta I_{PV} = 0$	No action

To solve these problems, adaptive hill-climbing, adaptive P&O, and adaptive INC algorithms using variable step sizes were introduced [25–28]. In each adaptive MPPT algorithm, a variable step size is automatically adjusted according to the slope formula consisting of the variations (ΔP_{PV} and ΔV_{PV}) and variable step coefficients (a, N, and M); $a{\cdot}\Delta P_{PV}/\Delta V_{PV}$ in [25,26], $N{\cdot}\Delta P_{PV}/\Delta V_{PV}$ in [27], and $M{\cdot}\Delta P_{PV}/\Delta V_{PV}$ in [28]. The variable step size becomes high far from the MPP and low near the MPP. Therefore, the adaptive MPPT algorithms can achieve fast tracking speed in the operating range far from the MPP and small oscillations in the operating range near the MPP. However, these algorithms still have oscillations at the MPP because the variables (ΔP_{PV} and ΔV_{PV}) in the slope formula are easily affected by sensing and calculation errors, sensing noise, and ripples of V_{PV} and I_{PV}.

In [29–32], the MPPT algorithms with intelligent prediction were introduced. Some of the algorithms used fuzzy logic to vary the step size [29–31]; the fuzzy logic method consists of fuzzification, a fuzzy rule-based lookup table (Table 3), and defuzzification. First, the input variables (ΔI_{PV} and ΔP_{PV}) are converted into fuzzy subsets such as negative big (NB), negative small (NS), zero (ZO), positive small (PS), and positive big (PB) depending on ΔI_{PV} and ΔP_{PV}. Then, one fuzzy subset is determined based on the rule-based lookup table and it provides a numeric step size. The MPPT algorithm using fuzzy logic is effective in dealing with the nonlinear characteristics of the PV module because the fuzzy logic divides the nonlinear system in the fuzzy subsets defined by the variables (ΔI_{PV} and ΔP_{PV}) and controls the nonlinearity of system using the rule base for each area (See Table 3). Therefore, it can track the MPP well under dynamic weather conditions. However, compared to other MPPT algorithms such as adaptive P&O and INC algorithms, it requires a digital signal processor (DSP) with higher specification and also makes the user's design more difficult because of its higher complexity of algorithm and execution process [27,31]. This is a disadvantage for commercialization. The other MPPT algorithm of [32] changed the step size by using a PID controller tuned by genetic algorithm. First, the genetic algorithm evaluates a number of solutions known as chromosomes using a fitness function, and later, genetic operators (selection, crossover, and mutation) are applied until a stop criterion is satisfied (Figure 4). Based on this principle of the genetic algorithm, the PID coefficients (K_p, K_i, and K_d) are determined to vary the step size. This MPPT algorithm also has good performance under dynamic weather conditions, but it requires high computational capability and sophisticated controllers due to its complexity.

Table 3. Rule-based lookup table based on the fuzzy logic.

ΔI_{PV}	ΔP_{PV}				
	NB	**NS**	**ZO**	**PS**	**PB**
NB	NB	NS	NS	ZO	ZO
NS	NS	ZO	ZO	ZO	PS
ZO	ZO	ZO	ZO	PS	PS
PS	ZO	PS	PS	PS	PB
PB	PS	PS	PB	PB	PB

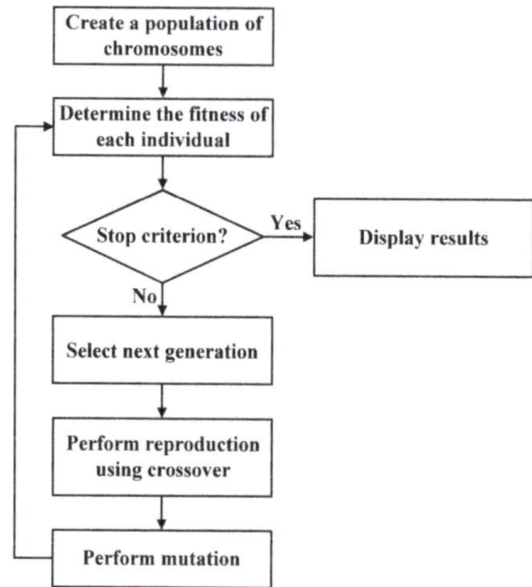

Figure 4. Flow chart of the genetic algorithm.

In this paper, an advanced MPPT algorithm for DMPPT in the MLPEs is proposed, which improves the tracking speed and accuracy under both steady and dynamic weather conditions. Similar to the adaptive MPPT algorithms, the proposed MPPT algorithm automatically changes the variable step size according to the slope of $\Delta P_{PV}/\Delta V_{PV}$ in the operating range far from the MPP. However, in the operating range near the MPP, it accurately tracks the MPP using a small fixed step size. As a result, the proposed MPPT algorithm achieves small oscillations at the MPP and a fast dynamic response. In addition, compared with the MPPT algorithms with intelligent control such as fuzzy logic and PID, the proposed MPPT algorithm can reduce the computational load of DSP because it is based on simple P&O method. Therefore, it allows manufacturers to use cheap DSPs for the PV system.

The distributed PV system and proposed MPPT algorithm are described in Section 2, the experimental results and discussion are presented in Section 3, and a conclusion is given in Section 4.

2. Distributed PV System and Proposed MPPT Algorithm

2.1. Distributed PV System

A distributed PV system with the MPPT algorithm is shown in Figure 5; the system consists of one PV module and one MLPE. Due to the fact that the PV module converts sunlight into low DC voltage, the MLPE requires a high voltage gain to convert low DC voltage into high AC voltage with grid frequency. The MPPT algorithm controls the duty ratio (D) of the MLPE to operate at the MPP of the PV module. To perform the MPPT algorithm, the voltage (V_{PV}) and current (I_{PV}) of the PV module are used as input signals.

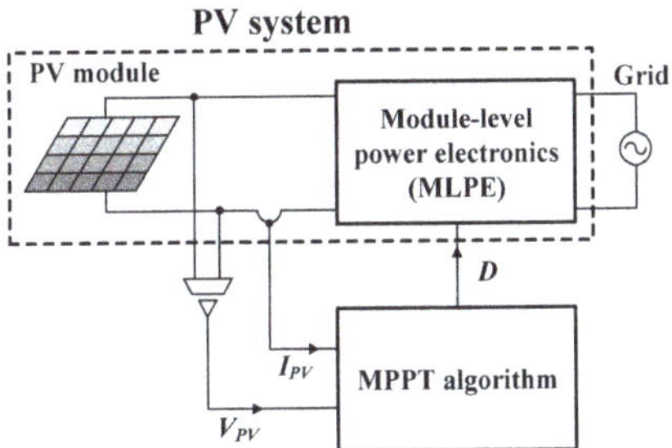

Figure 5. Diagram of the distributed photovoltaic (PV) system with maximum power point tracking (MPPT) algorithm.

2.1.1. PV Module

Based on the references [33,34], the equivalent circuit of the PV module can be derived as shown in Figure 6. The PV module consists of several PV cells and two parasitic resistances (R_S, R_P), where each PV cell has a combined structure of an ideal current source (I_P) and a diode (D_P). The current (I_{PV}) of the PV module can be derived as

$$I_{PV} = I_P - I_D - I_{Rp} \tag{1}$$

where $I_P = I_{P1} + I_{P2} \cdots + I_{Pn}$ and $I_D = I_{D1} + I_{D2} \cdots + I_{Dn}$. Assuming all PV cells are the same, $I_P = nI_{P1}$ and $I_D = nI_{D1}$ are obtained. Using the Shockley diode equation, I_D is given by

$$I_D = nI_0\left[e^{(V_{PV}+R_SI_{PV})/(an_SV_t)} - 1\right] \tag{2}$$

where n is the number of PV cells connected in parallel, n_s is the number of PV cells connected in series, I_0 is the saturation current of the diode, a is the diode ideality constant, and $V_t = kT/q$ is the thermal voltage of the diode with $q = 1.602 \times 10^{-19}$ C, $k = 1.381 \times 10^{-23}$ J/K, and T is an ambient temperature.

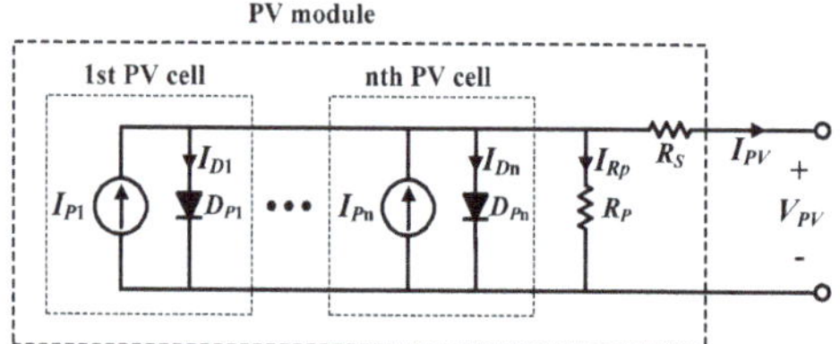

Figure 6. Equivalent circuit of the PV module.

As the voltage across R_P is given by $V_{PV} + R_S I_{PV}$, the current of R_P is obtained as

$$I_{Rp} = \frac{V_{PV} + R_S I_{PV}}{R_P}. \tag{3}$$

Then, inserting (2) and (3) into (1) results in

$$I_{PV} = I_P - nI_0\left[e^{(V_{PV}+R_S I_{PV})/(aV_t)} - 1\right] - \frac{V_{PV} + R_S I_{PV}}{R_P}. \tag{4}$$

Using (4), the characteristic curves of the PV module are drawn as shown in Figure 7. Figure 7a shows the current–voltage characteristic curve of the PV module and Figure 7b shows the power–voltage characteristic curve. These curves have notable variables: short-circuit current (I_{SC}), open-circuit voltage (V_{OC}), maximum power point (MPP), voltage at MPP (V_{MPP}), current at MPP (I_{MPP}), and power at MPP (P_{MPP}).

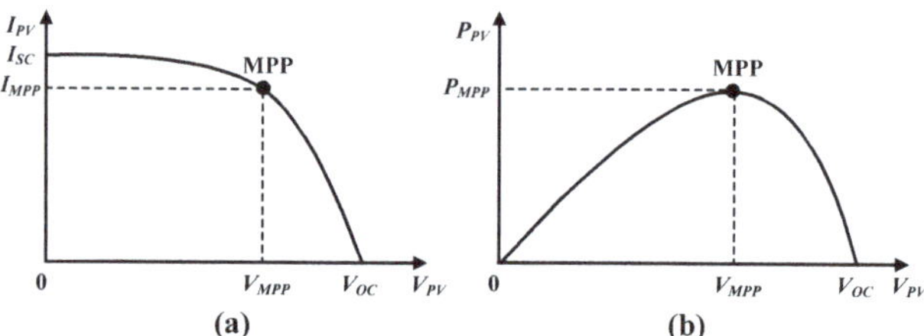

Figure 7. (**a**) Current–voltage characteristic curve and (**b**) power–voltage characteristic curve of the PV module.

The current of the PV cell depends on the solar irradiance and the temperature as follows;

$$I_P = n[I_{P,STC} + K_I(T - T_n)]\frac{G}{G_n} \tag{5}$$

where $I_{P,STC}$ is a light-generated current at the standard test condition (STC, 25°C and 1000 W/m²), K_I is a temperature coefficient, T_n is a nominal temperature (25°C), G is an irradiation level of the PV module, and G_n is a nominal irradiance level (1000 W/m²). Using (5), Equation (4) can be represented as

$$I_{PV} = n[I_{P,STC} + K_I(T - T_n)]\frac{G}{G_n} - nI_0\left[e^{(V_{PV}+R_S I_{PV})/(aV_t)} - 1\right] - \frac{V_{PV} + R_S I_{PV}}{R_P}. \tag{6}$$

Equation (6) shows that the characteristic curve of the PV module can be changed according to the solar irradiance level and the temperature.

2.1.2. Module-Level Power Electronics (MLPE)

As shown in Figure 8, the MLPE is classified into a micro-inverter and a DC-optimizer. The micro-inverter consists of a DC–DC converter with high voltage gain and a DC–AC inverter, and the DC-optimizer has only one DC–DC converter with high voltage gain. Therefore, the DC-optimizer can cost less than the micro-inverter, but it needs a DC–AC inverter with high power capability. In the micro-inverter and the DC-optimizer, one MPPT algorithm is applied to each DC–DC converter to optimize each PV module. Therefore, the MPPT algorithms used in the MLPEs are called a module-level MPPT, or DMPPT.

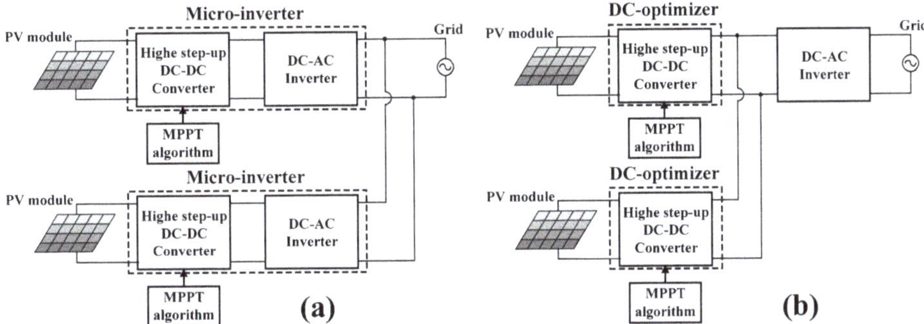

Figure 8. Diagrams of (**a**) the micro-inverter and (**b**) the direct current (DC)-optimizer.

2.2. Prospoed MPPT Algorithm

2.2.1. Principle of the Algorithm

The proposed MPPT algorithm uses two methods that can apply the optimal step size to each operating range. In the operating range far from the MPP (non-MPP region), a variable step size ($=k_1 \cdot S \cdot V_{step}$) is automatically adjusted according to the slope of $\Delta P_{PV}/\Delta V_{PV}$ for fast dynamic response. Here, k_1 is a constant coefficient for variable step size, S is a slope coefficient calculated as $|\Delta P_{PV}/\Delta V_{PV}|$, and V_{step} is a fixed step size. In the operating range near the MPP (MPP region), a small fixed step size ($=k_2 \cdot V_{step}$) is used to minimize the oscillations at the MPP, where k_2 is a constant coefficient for small fixed step size.

As shown in Figure 9, the proposed MPPT algorithm has two operating regions including the MPP ($P_{PV} > \beta \cdot P_{MPP}$) and non-MPP ($P_{PV} < \beta \cdot P_{MPP}$) regions, where β is an MPP region coefficient. The operating point starts at $V_{PV} = V_{OC}$ and move with fast tracking speed toward the MPP (V_{MPP}, P_{MPP}). The direction of the operating point is determined using ΔP_{PV} and ΔV_{PV}, where Δ means the difference between the present and previous values. In addition, the present P_{PV} is compared with P_{MPP} to search the MPP.

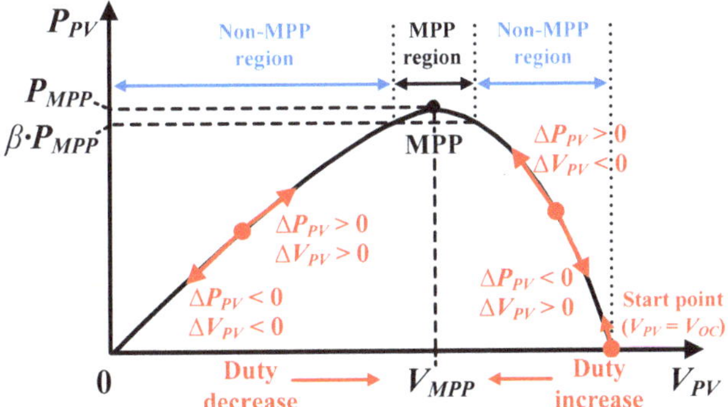

Figure 9. Principle of operation for the proposed MPPT algorithm.

2.2.2. Flow Chart

The flow chart of the proposed MPPT algorithm is shown in Figure 10, and it is described as follows:

1. The present V_{PV} and I_{PV} of the PV module are used as the input signals for the proposed MPPT algorithm. The variable "Flag_start" is preset to 1 for fast tracking speed at the starting point of $V_{PV} = V_{OC}$, and the variable "Flag_reset" is preset to 1 for setting the P_{MPP} and V_{MPP} to the present P_{PV} and V_{PV}, where $P_{PV} = V_{PV} \cdot I_{PV}$.

2. If Flag_start is 1, it is determined that the operating point is located at the starting point of $V_{PV} = V_{OC}$. Therefore, the reference variable (V_{ref}) is initially set to $1/V_{PV}$ ($=1/V_{OC}$) and the operating point moves rapidly toward the MPP. After that, Flag_start is set to 0.

3. In this process, P_{PV} is calculated as $V_{PV} \cdot I_{PV}$, ΔP_{PV} and ΔV_{PV} are calculated using present (P_{PV} and V_{PV}) and previous (P_{PV_b} and V_{PV_b}) values, and the slope coefficient (S) is calculated as $|\Delta P_{PV}/\Delta V_{PV}|$.

4. If Flag_reset is 1, P_{MPP} and V_{MPP} are set to the present P_{PV} and V_{PV}, and then Flag_reset is set to 0.

5. If the present P_{PV} is higher than P_{MPP}, it is determined that the MPP has not been found yet. Therefore, the operating point is forced to keep moving toward the MPP, and P_{MPP} and V_{MPP} are reset to the present P_{PV} and V_{PV}. To quickly find the MPP, the variable step size ($=k_1 \cdot S \cdot V_{step}$) is used in this process.

6. If the operating point is located in the MPP region, the small fixed step size ($=k_2 \cdot V_{step}$) is used to track the MPP accurately.

7. If P_{PV} is lower than the boundary value ($\beta \cdot P_{MPP}$) between the MPP and non-MPP regions, it is determined that the operating point is located in non-MPP region. This process is usually performed under dynamic weather conditions because the MPP changes under these conditions. Therefore, Flag_reset is set to 1 to find a new MPP, and the variable step size ($=k_1 \cdot S \cdot V_{step}$) is automatically adjusted according to the slope of $\Delta P_{PV}/\Delta V_{PV}$ for a fast dynamic response.

8. V_{ref} is limited by the maximum and minimum values ($V_{ref,max}$ and $V_{ref,min}$) of V_{ref} to prevent malfunction of the DC–DC converter in the MLPEs.

9. Through the above processes, a new V_{ref} is obtained. V_{ref} is compared with the carrier signal ($V_{carrier}$) in the digital signal processor (DSP), and a new duty ratio (D) is generated to control the DC–DC converter in the MLPE (Figure 11). In addition, the previous values (P_{PV_b} and V_{PV_b}) are obtained at this time.

The above processes from (1) to (9) are repeated to operate the DC–DC converter at the MPP of the PV module.

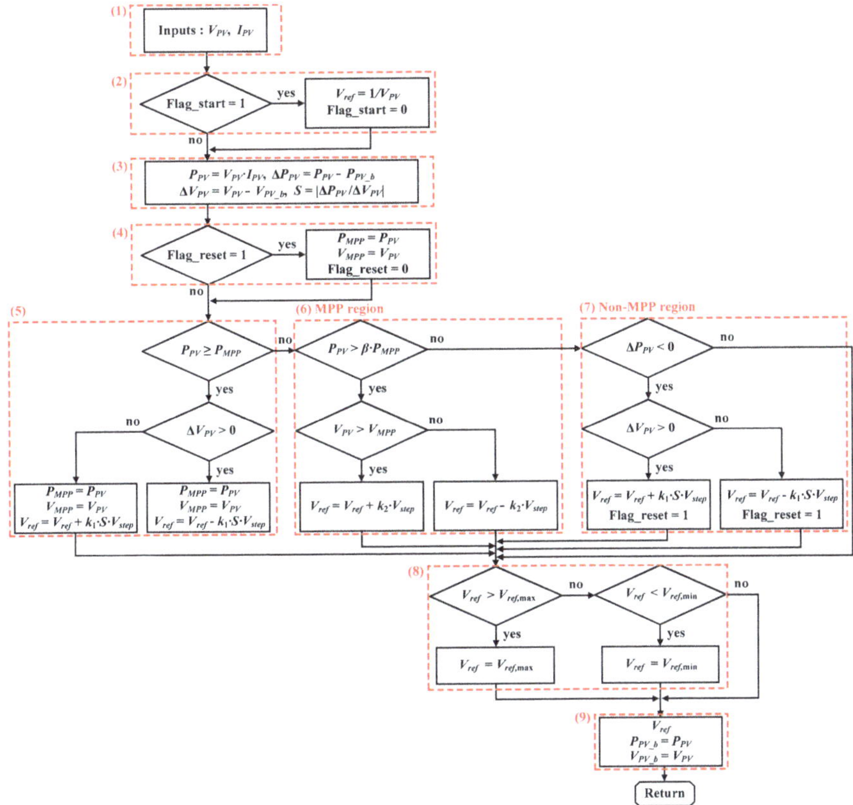

Figure 10. Flow chart of the proposed MPPT algorithm.

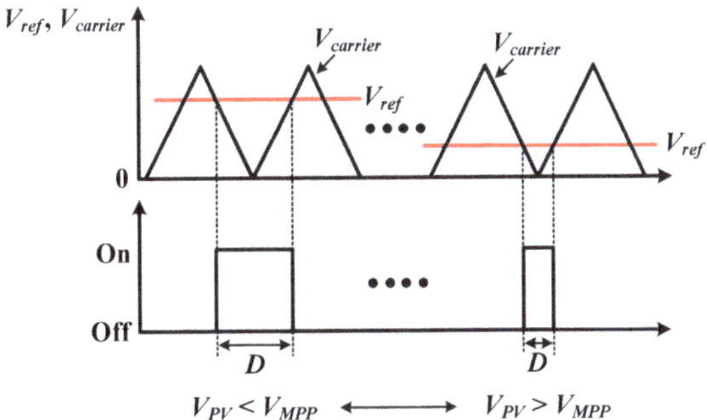

Figure 11. The process for generating a duty ratio (D) in the digital signal processor (DSP).

2.2.3. Design Considerations

(1) aximum ($V_{ref,max}$) and minimum ($V_{ref,min}$) of V_{ref}

The DC–DC converter in the MLPEs has an input voltage range of $V_{PV,min} \leq V_{PV} \leq V_{PV,max}$ due to its limited voltage gain. Since V_{ref} is calculated as $1/V_{PV}$, $V_{ref,max}$ and $V_{ref,min}$ can be defined as $1/V_{PV,min}$ and $1/V_{PV,max}$, respectively.

(2) Coefficient (k_1) for variable step size and coefficient (k_2) for small fixed step size

The proposed MPPT algorithm uses a variable step size ($=k_1{\cdot}S{\cdot}V_{step}$) for fast dynamic response in operating range far from the MPP and a small fixed step size ($=k_2{\cdot}V_{step}$) for small oscillations in operating range near the MPP, where k_1, k_2, and V_{step} have constant values. Due to the fact that the slope coefficient (S) is calculated as $|\Delta P_{PV}/\Delta V_{PV}|$, the variation of the reference variable (V_{ref}) is given by

$$\left|\Delta V_{ref}\right| = k_1 \left|\frac{\Delta P_{PV}}{\Delta V_{PV}}\right| V_{step} \tag{7}$$

where Δ means a difference between the present and previous values. The operating point of the proposed MPPT algorithm reaches the MPP ($V_{PV} = V_{MPP}$, $P_{PV} = P_{MPP}$) after starting at the point ($V_{PV} = V_{OC}$, $P_{PV} = 0$), where V_{OC} is an open-circuit voltage of the PV module. Considering only the starting point and the MPP at the STC (25 °C and 1000 W/m^2), the above equation can be represented using $V_{ref} = 1/V_{PV}$ as

$$\frac{1}{V_{MPP,STC}} - \frac{1}{V_{OC}} = k_1 \frac{P_{MPP,STC}}{V_{OC} - V_{MPP,STC}} V_{step} \tag{8}$$

where $V_{MPP,STC}$ and $P_{MPP,STC}$ are the voltage and power at the MPP under the STC (25 °C and 1000 W/m^2).

Therefore, k_1 can be obtained as

$$k_1 = \frac{(V_{OC} - V_{MPP,STC})^2}{V_{OC} V_{MPP,STC} V_{step} P_{MPP,STC}}. \tag{9}$$

The small step coefficient (k_2) can be determined by considering the resolution of V_{PV} in the operating range near the MPP. If accuracy of the third decimal point of V_{PV} is required in operating range near the MPP, $|\Delta V_{ref}| = k_2{\cdot}V_{step}$ can be represented using $V_{ref} = 1/V_{PV}$ as

$$\frac{1}{V_{MPP,STC}} - \frac{1}{V_{MPP,STC} + 0.001} = k_2 V_{step}. \tag{10}$$

Therefore, k_2 is calculated as

$$k_2 = \frac{0.001}{V_{MPP,STC}(V_{MPP,STC} + 0.001) V_{step}}. \tag{11}$$

(3)　MPP region coefficient (β)

$\beta{\cdot}P_{MPP}$ is a boundary value between the MPP and non-MPP regions. For fast dynamic response, β should be close to one; however, if it is too close to one, oscillations can occur at the boundary of $P_{PV} = \beta{\cdot}P_{MPP}$. Therefore, $\beta = 0.9{\sim}0.95$ is recommended.

(4)　Fixed step size (V_{step})

To optimize the proposed MPPT algorithm in a given DC–DC converter topology and operating conditions, V_{step} is determined after several experiments. Therefore, V_{step} has a user-defined value, and $V_{step} = 10$ was used for the experiments in this paper.

3. Experimental Results and Discussion

3.1. Experimental Results

Figure 12 shows the equipment settings to evaluate the performances of the MPPT algorithms. The output cable of the PV simulator (ETS60 from AMETEK Inc.) is connected to the input of the DC–DC converter (boost half-bridge topology), and the LAN cable of the PV simulator is connected to the notebook computer to set the characteristic curve (Figure 13) of the commercial 300-W PV module (Q.PEAK-G4.1 300 from Hanwha Inc.). Here, according to the irradiance level (E) and the module temperature (T), the electrical characteristics of the PV module "Q.PEAK-G4.1 300" are listed in Table 4. The output of the DC–DC converter is connected to the electronic load (DL1000H from NF Corp.) to consume the energy, and the notebook computer is connected to the DSP (TMS320F28335 from Texas Instruments Inc.) in the DC–DC converter by USB cable. The power meter (PW3336, from HIOKI Inc.) is used to measure the voltage (V_{PV}), current (I_{PV}), and power (P_{PV}) of the PV simulator. Based on the sensed V_{PV} and I_{PV}, the DSP runs the MPPT algorithm and generates the gate signals (v_{gs1}, v_{gs2}) for operating the DC–DC converter. To compare the dynamic and energy utilization performances of the MPPT algorithms, four experiments were performed as follows.

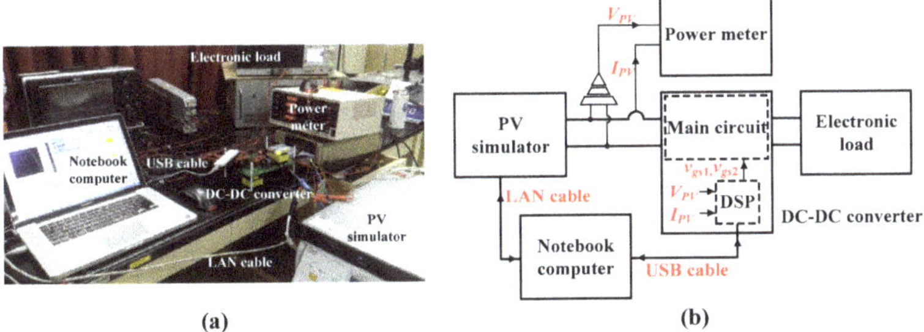

(a)   (b)

Figure 12. (**a**) Photograph and (**b**) block diagram of the equipment settings for experiments to evaluate the MPPT performance.

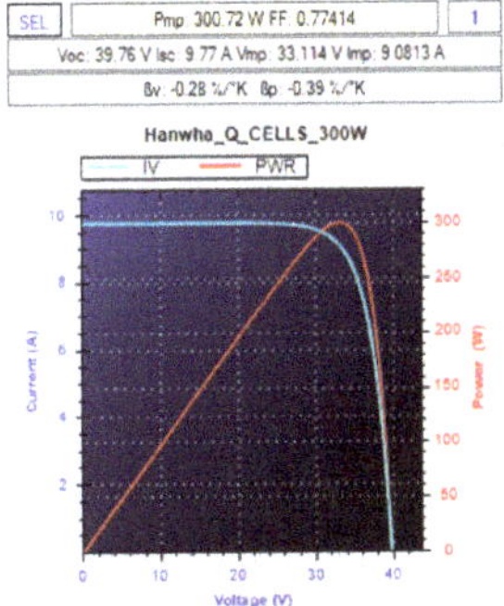

Figure 13. Electrical characteristic curve of a 300 W PV module "Q.PEAK-G4.1 300" under the standard test conditions (STC) of irradiance level of 1000 W/m^2 and module temperature of 25 °C.

Table 4. Electrical characteristics of the PV module "Q.PEAK-G4.1 300" at different irradiance levels (E) and module temperatures (T).

Conditions	Open-Circuit Voltage (V_{OC})	Short-Circuit Current (I_{SC})	Voltage at MPP (V_{MPP})	Current at MPP (I_{MPP})	Power at MPP (P_{MPP})
E = 1000 W/m^2 and T = 25 °C	39.76 V	9.77 A	33.11 V	9.082 A	300.71 W
E = 100 W/m^2 and T = 25 °C	35.78 V	1.086 A	29.8 V	1.009 A	30.07 W
E = 700 W/m^2 and T = 15 °C	40.24 V	7.021 A	33.51 V	6.527 A	218.72 W
E = 700 W/m^2 and T = 75 °C	33.66 V	6.502 A	28.04 V	6.044 A	169.47 W

First, the proposed MPPT algorithm and the conventional MPPT algorithms including a basic P&O of [21], an adaptive P&O of [26], and an adaptive INC of [27] were tested at the STC of E = 1000 W/m^2 and T = 25 °C. Figure 14 shows the measured V_{PV}, I_{PV}, and P_{PV} waveforms of the PV module when each MPPT algorithm is applied to a DC–DC converter. The basic P&O algorithm with a small fixed step size tracked the maximum power point (MPP) slowly, but it had small oscillations after reaching the MPP (Figure 14a). When the fixed step size increased, the tracking speed of the basic P&O algorithm increased as well, but the oscillations also increased (Figure 14b). To solve the problem of the fixed step size, the adaptive P&O and adaptive INC algorithms used a variable step size proportional to $k_1 \cdot \Delta P_{PV}/\Delta V_{PV}$. However, these algorithms still had the oscillations at the MPP because the $\Delta P_{PV}/\Delta V_{PV}$ is easily affected by sensing and calculation errors, sensing noise, and the ripples of V_{PV} and I_{PV} (Figure 14c–f). If the coefficient k_1 is smaller, the oscillations at the MPP decreases, but the tracking speed is slower. To improve the problems of the previous MPPT algorithms mentioned above, the proposed MPPT algorithm uses two methods that can apply the optimal step size to each operating range. In the operating range near the MPP, a small fixed step size is used to minimize the oscillations at the MPP, but in the operating range far from the MPP, a variable step size proportional to $k_1 \cdot \Delta P_{PV}/\Delta V_{PV}$ is used to achieve fast tracking speed. The proposed MPPT algorithm can adjust the tracking speed using k_1 so that it can track the MPP quickly. As a result, the proposed MPPT algorithm tracked the MPP with faster tracking speed and had smaller oscillations at the MPP when compared with the conventional algorithms (Figure 14g,h).

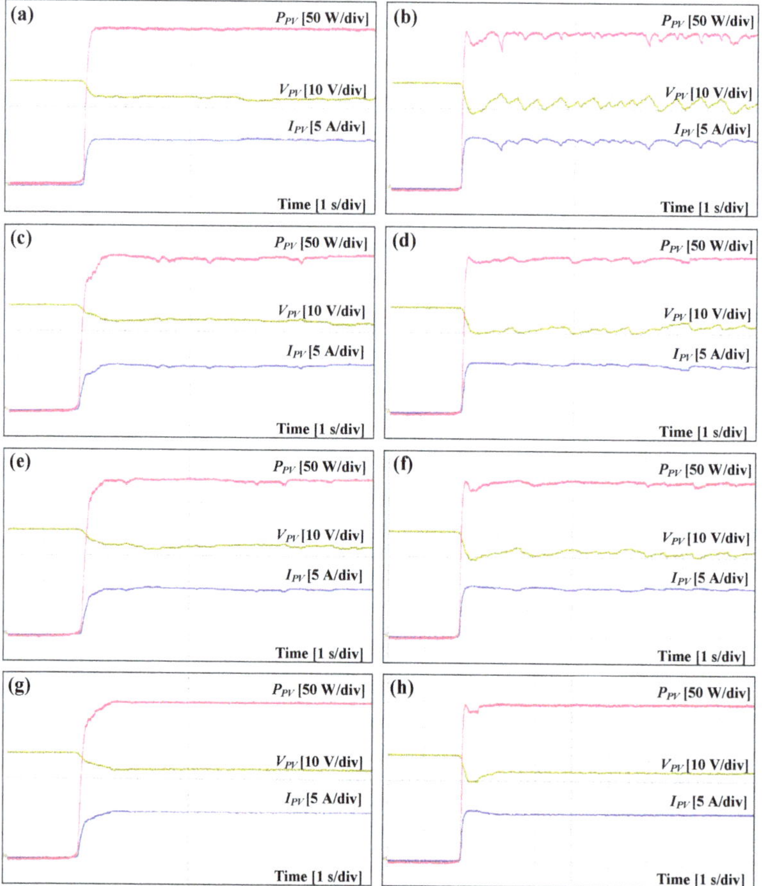

Figure 14. Voltage (V_{PV}), current (I_{PV}), and power (P_{PV}) waveforms of the PV module measured using the basic P&O algorithm with (**a**) small fixed step size or (**b**) large fixed step size, using the adaptive P&O algorithm with (**c**) low k_1(=0.00002) or (**d**) high k_1(=0.0001), using the adaptive INC algorithm with (**e**) low k_1(=0.00002) or (**f**) high k_1(=0.0001), and using the proposed MPPT algorithm with (**g**) low k_1(=0.00002) and small fixed step size or (**h**) high k_1(=0.0001) and small fixed step size.

The MPPT efficiencies of the proposed and conventional MPPT algorithms were measured for two minutes after reaching the MPP under the same conditions as in Figure 14 (Figure 15). When a large fixed step size was applied, the basic P&O algorithm had the lowest average MPPT efficiency of 97.8% due to the largest oscillations at the MPP (Figure 15a). The adaptive P&O and adaptive INC algorithms with high $k_1(=0.0001)$ had higher average MPPT efficiencies of 98.5% and 98.7% than that of the basic P&O algorithm because they used variable step sizes (Figure 15b,c). In addition, the adaptive P&O and adaptive INC algorithms had higher average MPPT efficiencies of 99.1% and 98.9% than 98.7% of the basic P&O algorithm when small fixed step size and low $k_1(=0.00002)$ were applied. The proposed MPPT algorithm had the highest average MPPT efficiencies of 99.4% and 99.7% at low $k_1(=0.00002)$ and high $k_1(=0.0001)$ because it had the smallest oscillations at the MPP (Figure 15d).

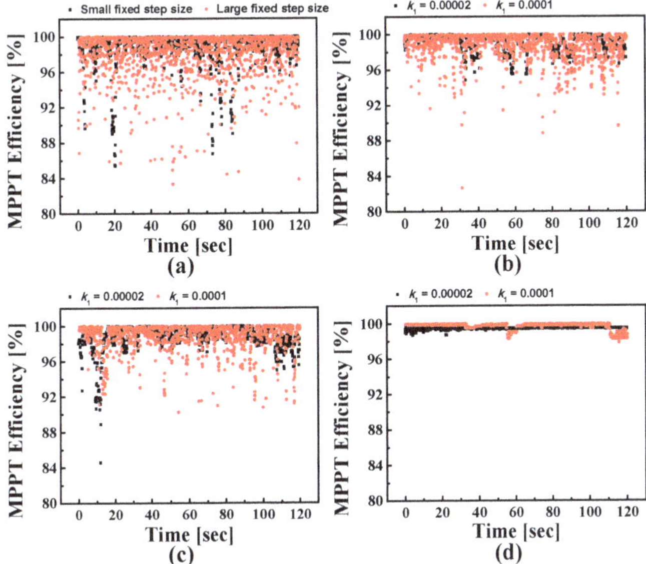

Figure 15. Measured MPPT efficiencies of (**a**) the basic P&O with small or large fixe step sizes, (**b**) the adaptive P&O with low $k_1(=0.00002)$ or high $k_1(=0.0001)$, (**c**) the adaptive INC with low $k_1(=0.00002)$ or high $k_1(=0.0001)$, and (**d**) the proposed MPPT algorithms with low $k_1(=0.00002)$ or high $k_1(=0.0001)$.

The P_{PV} waveforms were measured to compare the performances of the proposed and conventional MPPT algorithms when irradiance level changed abruptly from 1000 W/m^2 to 100 W/m^2 or from 100 W/m^2 to 1000 W/m^2 at a constant temperature of 25 °C (Figure 16). In each MPPT algorithm, the fixed step size and k_1 were optimized by considering the trade-off between the tracking speed and the oscillations under the given conditions. When irradiance level changed from 100 W/m^2 to 1000 W/m^2, the basic P&O algorithm tracked the MPP within 20 s but had large oscillations (Figure 16a). The adaptive P&O and adaptive INC algorithms also tracked the MPP within 20 s, but they still had oscillations (Figure 16b,c). As shown in Figure 14c–f, these algorithms could reduce the oscillations using a small fixed step size or k_1, but the tracking was slower. The proposed MPPT algorithm had no trade-off between the tracking speed and the oscillations because it uses a variable step size proportional to $k_1 \cdot \Delta P_{PV} / \Delta V_{PV}$ in the operating range far from the MPP and a small fixed step size in the operating range near the MPP. Therefore, when the irradiance level changed from 100 W/m^2 to 1000 W/m^2, the proposed MPPT algorithm showed smaller oscillations and faster tracking speed than other algorithms (Figure 16d).

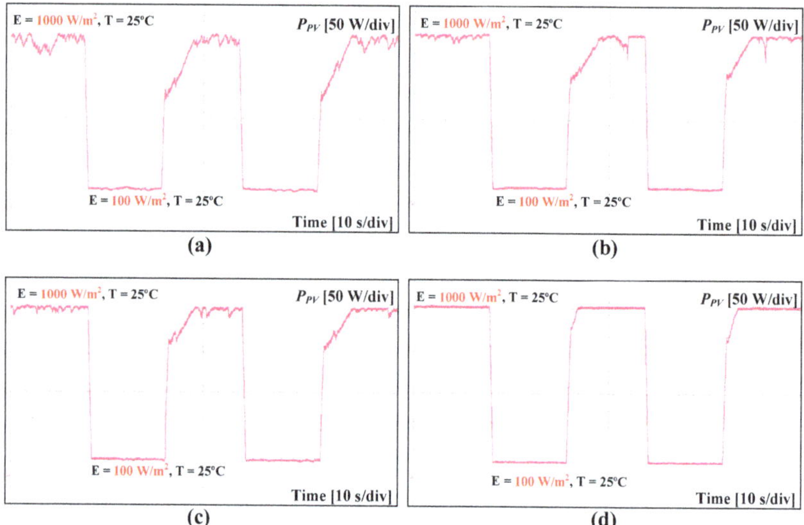

Figure 16. Power (P_{PV}) waveforms of the PV module measured using (**a**) the basic P&O algorithm, (**b**) the adaptive P&O algorithm, (**c**) the adaptive INC algorithm, and (**d**) the proposed MPPT algorithm under abruptly changing irradiance conditions.

Figure 17 shows the P_{PV} waveforms measured to compare the performances of the proposed and conventional MPPT algorithms when the temperature changed abruptly from 15 °C to 75 °C or from 75 °C to 15 °C at a constant irradiance of 700 W/m². In each MPPT algorithm, the fixed step size and k_1 were optimized by considering the trade-off between the tracking speed and the oscillations under the given conditions. The basic P&O algorithm slowly tracked the MPP due to the fixed step size when the temperature changed from 75 °C to 15 °C (Figure 17a). The adaptive P&O and adaptive INC algorithms tracked the MPP faster than the basic P&O, but they still had oscillations after reaching the MPP (Figure 17b,c). The proposed MPPT algorithm not only had the fastest tracking speed but also had the smallest oscillations after reaching the MPP because it had no trade-off between the tracking speed and the oscillations (Figure 17d). The above results of the comparison experiments are summarized in Table 5.

3.2. Discussion

Among many MPPT algorithms, the P&O method was widely used due to its simple principle and ease of implementation. However, the basic P&O algorithm of [21] had a trade-off between tracking speed and oscillations due to a fixed step size (Figure 14a,b). To solve this problem, the adaptive P&O and adaptive INC algorithms using a variable step size were introduced [26,27]. These adaptive MPPT algorithms can reduce the oscillations at the MPP because the variable step size is automatically adjusted according to the slope ($\Delta P_{PV}/\Delta V_{PV}$) of the P–V curve (Figure 14c–f). However, the adaptive MPPT algorithms still had oscillations after reaching the MPP because the calculated $\Delta P_{PV}/\Delta V_{PV}$ is easily affected by sensing and calculation errors, sensing noise, and the ripples of V_{PV} and I_{PV}. To improve the performances of the conventional MPPT algorithms mentioned above, the proposed MPPT algorithm used a small fixed step size in operating range near the MPP and a variable step size proportional to $k_1 \cdot \Delta P_{PV}/\Delta V_{PV}$ in operating range far from the MPP. As a result, the proposed MPPT algorithm had higher MPPT efficiency than the conventional MPPT algorithms (Figure 15) and showed faster tracking speed and smaller oscillations under dynamic weather conditions (Figures 16 and 17). These advantages of the proposed MPPT algorithm enable the distributed PV system to operate at its maximum performance, regardless of weather conditions.

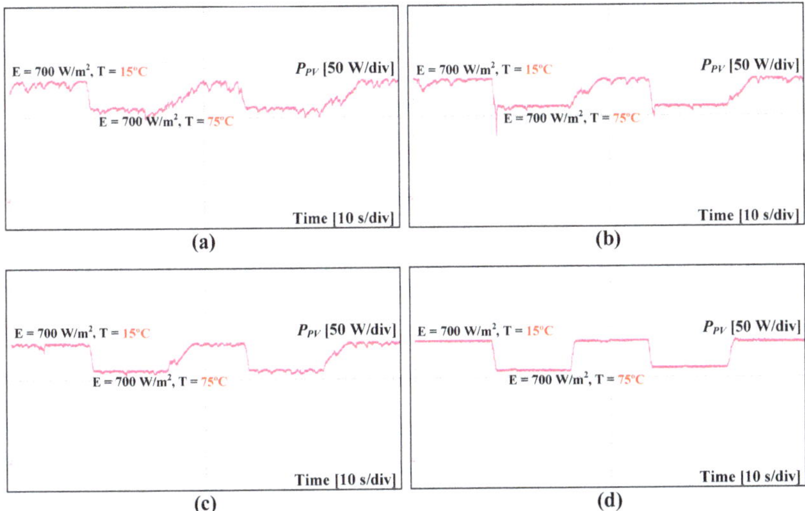

Figure 17. Power (P_{PV}) waveforms of the PV module measured using (**a**) the basic P&O algorithm, (**b**) the adaptive P&O algorithm, (**c**) the adaptive INC algorithm, and (**d**) the proposed MPPT algorithm under abruptly changing temperature conditions.

Table 5. Comparisons between the proposed and conventional MPPT algorithms.

Performance Parameters	Basic P&O Algorithm of [21]	Adaptive P&O Algorithm of [26]	Adaptive INC Algorithm of [27]	Proposed Algorithm
Implementation complexity	simple	medium	medium	medium
MPPT method	fixed step size ($k_2 V_{step}$) in whole operating range	variable step size ($k_1 V_{step} \Delta P_{PV}/\Delta V_{PV}$) in whole operating range	variable step size ($k_1 V_{step} \Delta P_{PV}/\Delta V_{PV}$) in whole operating range	small fixed step size ($k_2 V_{step}$) near the MPP, variable step size ($k_1 V_{step} \Delta P_{PV}/\Delta V_{PV}$) far from the MPP
MPPT efficiency	97.8%	98.5%	98.7%	99.7%
Performance at rapid change of irradiance	poor	medium	medium	good
Performance at rapid change of temperature	poor	medium	medium	good
Speed	fast	fast	fast	fast
Accuracy	low	medium	medium	high

4. Conclusions

In this paper, an advanced MPPT algorithm for distributed PV systems was proposed. The proposed MPPT algorithm improved the MPPT accuracy and dynamic response of the conventional MPPT algorithms by using two methods that can apply the optimal step size to each operating range. In the operating range near the MPP, a small fixed step size is used to minimize the oscillations at the MPP, but in the operating range far from the MPP, a variable step size proportional to $k_1 \cdot \Delta P_{PV}/\Delta V_{PV}$ is used to achieve fast tracking speed. These advantages of the proposed MPPT algorithm were verified by comparison with the conventional MPPT algorithms: a basic P&O algorithm with fixed step size, an adaptive P&O algorithm with variable step size, and an adaptive INC algorithm with a variable step size. In the experimental results, the proposed MPPT algorithm had the highest MPPT efficiency of 99.7% compared with the conventional MPPT algorithms. In addition, it showed the fastest tracking speed and smallest oscillations under abruptly changing irradiance levels and temperature conditions. Due to these advantages of the proposed MPPT algorithm, the PV systems using the proposed MPPT algorithm can produce more electrical energy than that using the conventional MPPT algorithms. Moreover, the proposed MPPT algorithm requires low computational load of DSP because the formulas used for step size are simple, which is a good advantage in terms of user convenience and cost that are important for commercialization. As a result, the proposed MPPT algorithm is well-suited for distributed PV systems requiring high MPPT efficiency and fast dynamic response.

Author Contributions: Both authors contributed equally to this work. H.-S.L. presented the main idea of the MPPT algorithm. H.-S.L. and J.-J.Y. analyzed the proposed MPPT algorithm and performed experiments. J.-J.Y. contributed to the overall composition and writing of the manuscript.

Funding: This work was supported by the National Research Foundation of Korea (NRF) grant funded by the government of Korea (MSIP) (NRF-2018R1A1A1A05079496). This research was also supported by the Korea Electric Power Corporation (Grant number: R17XA05-15).

Conflicts of Interest: The authors have no conflict of interest.

References

1. Malinowski, M.; Leon, J.I.; Abu-Rub, H. Solar photovoltaic and thermal energy systems: Current technology and future trends. *Proc. IEEE* **2017**, *105*, 2132–2146. [CrossRef]
2. Kroposki, B.; Johnson, B.; Zhang, Y.; Gevorgian, V.; Denholm, P.; Hodge, B.M.; Hannegan, B. Achieving a 100% renewable grid: Operating electric power systems with extremely high levels of variable renewable energy. *IEEE Power Energy Mag.* **2017**, *15*, 61–73. [CrossRef]
3. García-Rodríguez, L. Renewable Energy Applications in Desalination: State of the Art. *Sol. Energy* **2003**, *75*, 381–393. [CrossRef]
4. Renewables 2019 Global Status Report, Renewable Energy Policy Network for the 21st Century (REN21). 2019. Available online: https://www.ren21.net/gsr-2019/ (accessed on 2 September 2019).
5. Romero-Cadaval, E.; Francois, B.; Malinowski, M.; Zhong, Q.C. Grid-connected photovoltaic plants: An alternative energy source, replacing conventional sources. *IEEE Ind. Electron. Mag.* **2015**, *9*, 18–32. [CrossRef]
6. Zhao, P.; Suryanarayanan, S.; Simoes, M.G. An energy management system for building structures using a multi-agent decision-making control methodology. *IEEE Trans. Ind. Appl.* **2013**, *49*, 322–330. [CrossRef]
7. Strunz, K.; Abbasi, E.; Huu, D.N. DC microgrid for wind and solar power integration. *IEEE J. Emerg. Sel. Top. Power Electron.* **2014**, *2*, 115–126. [CrossRef]
8. Feldman, D.; Barbose, G.; Margolis, R.; Bolinger, M.; Chung, D.; Fu, R.; Seel, J.; Davidson, C.; Darghouth, N.; Wiser, R. Photovoltaic System Pricing Trends Historical, Recent, and Near-Term Projections 2015 Edition. Available online: https://escholarship.org/content/qt9pc3x32t/qt9pc3x32t.pdf (accessed on 2 September 2019).
9. Zsiborács, H.; Hegedűsné Baranyai, N.; Csányi, S.; Vincze, A.; Pintér, G. Economic Analysis of Grid-Connected PV System Regulations: A Hungarian Case Study. *Electronics* **2019**, *8*, 149. [CrossRef]
10. Camera, F.L. *Renewable Power Generation Costs in 2018*; International Renewable Energy Agency (IRENA): Abu Dhabi, UAE, 2019. Available online: https://www.irena.org/publications/2019/May/Renewable-power-generation-costs-in-2018 (accessed on 2 September 2019).
11. Meza, C.; Negroni, J.J.; Biel, D.; Guinjoan, F. Energy-balance modeling and discrete control for single-phase grid-connected PV central inverters. *IEEE Trans. Ind. Electron.* **2008**, *55*, 2734–2743. [CrossRef]
12. Flicker, J.; Tamizhmani, G.; Moorthy, M.K.; Thiagarajan, R.; Ayyanar, R. Accelerated testing of module-level power electronics for longterm reliability. *IEEE J. Photovolt.* **2017**, *7*, 259–267. [CrossRef]
13. Liu, Y.H.; Chen, J.H.; Huang, J.W. A review of maximum power point tracking techniques for use in partially shaded conditions. *Renew. Sustain. Energy Rev.* **2015**, *41*, 436–453. [CrossRef]
14. Gu, B.; Dominic, J.; Lai, J.S.; Zhao, Z.; Liu, C. High boost ratio hybrid transformer DC-DC converter for photovoltaic module applications. *IEEE Trans. Power Electron.* **2013**, *28*, 2048–2058. [CrossRef]

15. Sinapis, K.; Tzikas, C.; Litjens, G.; Van den Donker, M.; Folkerts, W.; Van Sark, W.G.J.H.M.; Smets, A. A comprehensive study on partial shading response of c-Si modules and yield modeling of string inverter and module level power electronics. *Sol. Energy* **2016**, *135*, 731–741. [CrossRef]
16. Aganza-Torres, A.; Cardenas, V.; Pacas, M.; Gonzales, M. An efficiency comparative analysis of isolated multi-source grid-connected PV generation systems based on a HF-link micro-inverter approach. *Sol. Energy* **2016**, *127*, 239–249. [CrossRef]
17. Luo, H.; Wen, H.; Li, X.; Jiang, L.; Hu, Y. Synchronous buck converter based low-cost and high-efficiency sub-module DMPPT PV system under partial shading conditions. *Energy Convers. Manag.* **2016**, *126*, 473–487. [CrossRef]
18. Femia, N.; Lisi, G.; Petrone, G.; Spagnuolo, G.; Vitelli, M. Distributed maximum power point tracking of photovoltaic arrays: Novel approach and system analysis. *IEEE Trans. Ind. Electron.* **2008**, *55*, 2610–2621. [CrossRef]
19. Wang, F.; Fang, Z.; Lee, F.C.; Zhu, T.; Yi, H. Analysis of existence-judging criteria for optimal power regions in DMPPT PV systems. *IEEE Trans. Energy Convers.* **2016**, *31*, 1433–1441. [CrossRef]
20. Ahmed, A.; Ran, L.; Bumby, J. Perturbation parameters design for hill climbing MPPT techniques. In Proceedings of the IEEE International Symposium on Industrial Electronics, Hangzhou, China, 28–31 May 2012. [CrossRef]
21. Hua, C.; Lin, J.; Shen, C. Implementation of a DSP-controlled photovoltaic system with peak power tracking. *IEEE Trans. Ind. Electron.* **1998**, *45*, 99–107. [CrossRef]
22. Veerachary, M.; Senjyu, T.; Uezato, K. Voltage-based maximum power point tracking control of PV system. *IEEE Trans. Aerosp. Eelctron. Syst.* **2002**, *38*, 262–270. [CrossRef]
23. Femia, N.; Petrone, G.; Spagnuolo, G.; Vitelli, M. Optimization of perturb and observe maximum power point tracking method. *IEEE Trans. Power Electron.* **2005**, *20*, 963–973. [CrossRef]
24. Safari, A.; Mekhilef, S. Simulation and hardware implementation of incremental conductance MPPT with direct control method using cuk converter. *IEEE Trans. Ind. Electron.* **2011**, *58*, 1154–1161. [CrossRef]
25. Xiao, W.; Dunford, W.G. A modified adaptive hill climbing MPPT method for photovoltaic power systems. In Proceedings of the IEEE 35th Annual Power Electronics Specialists Conference, Aachen, Germany, 20–25 June 2004. [CrossRef]
26. Zhang, X.; Zhang, H.; Zhang, H.; Zhang, P.; Wang, F.; Jia, H.; Song, D. A variable step-size P&O method in the application of MPPT control for a PV system. In Proceedings of the IEEE Advanced Information Management, Communicates, Electronic and Automation Control Conference (IMCEC), Xi'an, China, 3–5 October 2016. [CrossRef]
27. Liu, F.; Duan, S.; Liu, F.; Liu, B.; Kang, Y. A variable step size INC MPPT method for PV systems. *IEEE Trans. Ind. Electron.* **2008**, *55*, 2622–2628. [CrossRef]
28. Pandey, A.; Dasgupta, N.; Mukerjee, A.K. High-performance algorithms for drift avoidance and fast tracking in solar MPPT system. *IEEE Trans. Energy Convers.* **2008**, *23*, 681–689. [CrossRef]
29. Aashoor, F.A.O.; Robinson, F.V.P. A Variable Step Size Perturb and Observe Algorithm for Photovoltaic Maximum Power Point Tracking. In Proceedings of the 47th International Universities Power Engineering Conference (UPEC), London, UK, 4–7 September 2012. [CrossRef]
30. Macaulay, J.; Zhou, Z. A fuzzy logical-based variable step size P&O MPPT algorithm for photovoltaic system. *Energies* **2018**, *11*, 1340. [CrossRef]
31. Chen, Y.T.; Jhang, Y.C.; Liang, R.H. A fuzzy-logic based auto-scaling variable step-size MPPT method for PV systems. *Sol. Energy* **2016**, *126*, 53–63. [CrossRef]
32. Harrag, A.; Messalti, S. Variable step size modified p&o mppt algorithm using ga-based hybrid offline/online pid controller. *Renew. Sustain. Energy Rev.* **2015**, *49*, 1247–1260. [CrossRef]
33. Villalva, M.G.; Gazoli, J.R.; Filho, E.R. Comprehensive approach to modeling and simulation of photovoltaic arrays. *IEEE Trans. Power Electron.* **2009**, *24*, 1198–1208. [CrossRef]
34. Shongwe, S.; Hanif, M. Comparative analysis of different singlediode PV modeling methods. *IEEE J. Photovolt.* **2015**, *5*, 938–946. [CrossRef]

MDPI

St. Alban-Anlage 66

4052 Basel

Switzerland

Tel. +41 61 683 77 34

Fax +41 61 302 89 18

www.mdpi.com

Energies Editorial Office

E-mail: energies@mdpi.com

www.mdpi.com/journal/energies